# CHINESE
# RURAL
# DEVELOPMENT

**An East Gate Book**

# CHINESE RURAL DEVELOPMENT

## THE GREAT TRANSFORMATION

# William L. Parish, editor

*M.E. Sharpe, Inc.*

ARMONK, NEW YORK
LONDON, ENGLAND

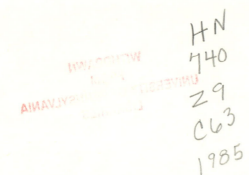

**East Gate Books** are edited by Douglas Merwin,
120 Buena Vista Drive, White Plains, New York 10603

©1985 by M. E. Sharpe, Inc.
80 Business Park Drive, Armonk, New York 10504

Available in the United Kingdom and Europe from M. E. Sharpe,
Publishers, 3 Henrietta Street, London WC2E 8LU.

**Library of Congress Cataloging in Publication Data**
Main entry under title:

Chinese rural development.


    Bibliography: p.
    1. Rural development—China—Addresses, essays, lectures. 2. China—
Rural conditions—Addresses, essays, lectures. 3. Peasantry—China—
Addresses, essays, lectures. I. Parish, William L.
HN740.Z9C63   1985       307′.14′0951        84-22193
ISBN 0-87332-314-9
ISBN 0-87332-344-0 (pbk.)

Printed in the United States of America

# Contents

## PART THREE

## New Patterns of Equality and Inequality

# Acknowledgments

This book stems from a 1981 conference on rural development funded by the Social Science Research Council's Joint Committee on Contemporary China. At that time Thomas Bernstein, Robert Dernberger, James Millar, Tang Tsou, Martin Whyte, and Edwin Winckler provided valuable critical comments that helped in the revision of the papers for this volume. In subsequent, updated versions of the essays, both Robert Dernberger and Michel Oksenberg provided additional suggestions for improving style and content. We only regret that we could not do more to live up to their high expectations.

In the initial stages of this project, Sophie Sa of the Social Science Research Council provided extremely helpful staff support, seeing that both the conference and the subsequent revisions of the papers moved along smoothly. We are grateful to all these people in helping shepherd this volume to completion.

W.L.P.

# CHINESE
# RURAL
# DEVELOPMENT

# 1

# Introduction: Historical Background and Current Issues

## William L. Parish

In the late 1970s and early 1980s, Chinese agriculture underwent a remarkable transformation. Peasants were given many new incentives to increase production, ranging from increased prices for their products, to more freedom to plant what they themselves found most profitable, to a new emphasis on private, family farming instead of joint, collective farming.

With these changes came a tremendous spurt in production and income. In the four years following 1978, peasant income about doubled. This was in marked contrast to the modest growth rate of the previous thirty years. Equally important, many peasants were lifted out of poverty. In 1978, one-third of all peasants had per capita incomes below one hundred yuan, but four years later only 3 percent had incomes this low.[1] These changes have been celebrated in the Chinese press and have become increasingly well known around the world.

Not all the trends were favorable. Abandoning much of the collective organization of agriculture that had been adopted in the mid-1950s weakened some of the social and economic services provided in the intervening twenty years. In some places, the new emphasis on family farming reduced the amount of labor and funds available for public infrastructure activities such as larger waterworks and larger agricultural machines. Social services were also weakened, with fewer villages supporting cooperative medical services or local schools. Income disparities increased in some villages, leading to jealousy of those families who did better and sometimes expropriation of their new-found riches until outside governmental bodies stepped in. Birth control became problematic as peasants calculated that they could prosper with more sons even if denied the increasingly less important collective rations. With fewer collective claims over individuals, illicit migration to cities became more difficult to control. These sorts of dysfunctions, even if not widespread, provided potential ammunition for domestic critics of the new policies.

It is difficult to disentangle the positive and negative consequences of the last few years. Needless to say, the official press tends to mention the negative consequences only obliquely while trumpeting loudly the many positive consequences. It is also difficult to disentangle the causes of changes that are fully reported in the press. Because so many changes in policy were introduced so rapidly, ranging from price changes to more individual family farming, weighing the influence of each policy is difficult.

Fortunately, we can get around some of these difficulties because of the remarkable openness of Chinese villages to foreign scholars during the initial years of these policy changes. This openness included not only a new outpouring of statistics on villages but also permission for a limited number of scholars to live and conduct research in villages. The results of much of this unusual research are captured in this volume.

The scholars writing here do emphasize many of the positive effects of the focus on private, family types of farming (part 2). But they also note the many other kinds of policy changes that were necessary and will continue to be necessary for increased prosperity in Chinese agriculture (part 1). Finally, they note the positive as well as negative social consequences that flowed from earlier collective policies (part 3). In short, written by scholars with firsthand experience in Chinese villages and with new Chinese statistics, this volume provides a comprehensive assessment of where Chinese collective agriculture has been, the forces that led to recent changes, and the problems that Chinese agriculture will continue to face in the future.

## Historical Developments

In the enthusiasm for rapid agricultural development since 1978, there is a tendency to forget the progress made in earlier years. In comparative, historical perspective, China's former experiment with collective agriculture may well continue to be judged as relatively successful. As new collectives were formed in the middle 1950s, many of the earlier mistakes of collective farming in the Soviet Union were avoided, and production continued to rise. This occurred in part because of the stepwise mobilization techniques that had been learned in the pre-1949 revolutionary war—peasants were not mobilized immediately into collectives but were taken first through land reform, small mutual aid teams, and then larger and larger collectives. Because the revolution had been a rural one, there was an ample rural leadership to help carry out these changes.

Perhaps most important, certain compromises were made with the existing natural social order in the countryside. Except for a three-year interregnum in the 1958–1961 Great Leap Forward, the basic production and income-sharing unit remained a group of twenty-five to thirty neighbors in the same village. Often these neighbors were close kinsmen as well. They were led not by an outsider sent in by the state but by a fellow neighbor who was paid out of their own farm receipts. Collective farmers kept 5–10 percent of the land for private vegetable plots, as well as family pigs and other sources of private income, totalling about one-fourth of all rural income. They also kept their own houses. More radical leaders were never very happy with these compromises, and there was frequent pressure to eliminate them by increasing the size of the collective unit, increasing collective pig production, limiting private income, and so forth. But the compromise more or less stuck through the 1970s, causing Chinese collective agriculture to be more family- and community-like than otherwise, and perhaps making it more palatable than in some other societies.

These compromises and the extensive rural administrative network helped produce significant progress on some dimensions. By standard development indicators, dramatic progress was made in the first twenty years of Chinese socialist agriculture. This was particularly true in the creation of an infrastructure that would support further development. A whole panoply of marketing and supply coops, agricultural machinery stations, hardware stores, repair shops, banks, credit coops, schools, hospitals, and clinics was created to serve peasants. In this volume, Marc Blecher describes many of these institutions in one particularly well-developed county. Also, statistics on linkages with the outside world help suggest the degree to which the larger state had entered the countryside. By 1975, virtually every rural brigade had a telephone. There were more than two wired loudspeakers for every team. Roads motorable by bus and truck extended deep into the countryside (table 1.1, panels A and E). Other statistics give a similar indication of progress. By 1979, for example, 87 percent of all commune seats and 63 percent of all brigades had electricity.

There was similar progress in the health and education fields (table 1.1, panel B). Between 1965 and 1975, with more emphasis on rural health care, the number of medical personnel and hospital beds increased rapidly. The number of university-trained Western-style medical doctors stagnated between 1965 and 1975, but the training of all other personnel including secondary-school-trained Western-style doctors, traditional-style Chinese doctors, and nurses as well as barefoot

Table 1.1

## Social Change Indicators by Year

| | 1952 | 1957 | 1965 | 1975 | 1982 |
|---|---|---|---|---|---|
| A. External Linkages | | | | | |
| Rural telephones ('000) | 58 | 200 | 492 | 659 | 804 |
| Wired broadcasting: | | | | | |
| —stations | 11 | 1,698 | 2,635 | 2,481 | 2,631 |
| —loudspeakers ('000,000) | — | 0.9 | 9 | 108 | 91† |
| Roads ('000 kms.) | 127 | 255 | 514 | 784 | 907 |
| B. Health and Education | | | | | |
| Rural hospital beds ('000) | 39 | 74 | 308 | 1,140‡ | 1,214† |
| Doctors and nurses ('000) | 690 | 1,039 | 1,532 | 2,057 | 3,143 |
| Barefoot paramedics ('000) | — | — | 94 | 1,559 | 1,349 |
| Rural deaths per 1,000 population | — | 11 | 10 | 8 | 7 |
| Primary pupils ('000,000) | 51 | 64 | 116 | 150 | 140 |
| Lower-middle pupils ('000,000) | 0.8* | 5 | 8 | 44 | 39 |
| Adult literacy (%) | — | — | 62§ | — | 76 |

| | 1952 | 1957 | 1965 | 1975 | 1982 |
|---|---|---|---|---|---|
| C. Income and Consumption | | | | | |
| Rural per capita consumption: | | | | | |
| —index (based on constant yuan) | 100 | 117 | 125 | 143 | 184† |
| —annual growth for period (%) | — | 3.2 | 0.8 | 1.4 | 5.2† |
| Output per capita (in kilograms): | | | | | |
| —grain | 288 | 306 | 272 | 312 | 350 |
| —edible oil | 4.4 | 6.6 | 5.1 | 5.0 | 11.7 |
| —fish | 2.9 | 4.9 | 4.2 | 4.8 | 5.1 |
| —meat | 6.0 | 6.2 | 7.7 | 8.7 | 13.4 |
| D. Agricultural Sidelines— | | | | | |
| % of total value of agricultural output | 4.4 | 4.3 | 6.5 | 9.1 | 16.0 |
| E. Administrative Units: | | | | | |
| Communes ('000) | — | 24# | 75 | 53 | 54 |
| Brigades ('000) | — | — | 648 | 677 | 719 |
| Teams ('000) | — | — | 5,412 | 4,826 | 5,977 |

*Sources: Zhongguo jingji nianjian*, 1981 edition, pp. VI-6, 9, 10, 12, 25; 1982 edition, pp. V-289, VIII-9, 11, 21, 28, 30, 31. Education statistics mostly from *Zhongguo baike nianjian*, 1980, p. 536; "1982 Census Results," *Beijing Review* 45 (1982): 20; *Zhongguo tongji nianjian*, 1983, pp. 105, 147, 151, 184, 320, 512, 534, 543, 545.

*Notes:* Some figures are for alternate years, as indicated by *1949, #1958, §1964, ‡1978, †1981.

paramedics continued at a fast pace. As a result of this training, China came to have more professional doctors, nurses, and hospital beds than virtually any country near its level of economic development.[2] With barefoot paramedics and others helping to educate villagers about sanitation, the death rate declined steadily. Thus, compared both to their own past and to farmers in other developing societies, Chinese farmers were more likely to live long lives in which they would be healthy, full-time laborers.

Not only were farmers healthier, they were also much better educated and prepared to absorb new technological changes that might be introduced in agriculture. By the mid-1970s 93 percent of all school-age children were said to be in school. And because of a push to create new two-year lower-middle schools, many were going on for a year or two beyond the five-year primary school as well. As a result of this educational expansion, over 70 percent of all adults were literate by the mid-1970s. As with health, this level of education is unusually high for a country at China's level of economic development.[3]

Despite the very real advances in communication, transportation, health, education, and other areas, there was considerable malaise in villages in the mid-1970s. One reason was that although income in 1975 was higher than two decades earlier, the progress over time had been rather slow (table 1.1, panel C). After the Great Leap, Chinese farmers did not return to 1957 consumption levels until almost 1965. Income growth in the late 1960s was respectable, but in the 1970s income again stagnated—the per capita income distributed by the collective was about the same in 1977 as it had been in 1971.[4] Consumption of many basic commodities stagnated as well. Grain availability per capita was only about what it had been two decades earlier, and the same was true of fish and other aquatic products. There was less cooking oil than before, for oil-bearing crops had fallen before the onslaught of the self-sufficiency, grain-first policy of the 1970s. The increasing supply of pork, beef, and mutton only partially compensated for the failure of these other goods to increase. In addition, as Mark Selden and Nick Lardy explain in this volume, significant poverty pockets remained. In 1977, almost one-fourth of China's 2,100 counties received per capita incomes below the poverty level of fifty yuan per capita.

There was little chance to mend these sorts of lingering problems until the moderate leaders who had taken over after Mao Zedong's death in 1976 consolidated their power. But when they did begin to see to these problems, they did so in a sweeping manner. In a set of new policies ratified at the December 1978 Third Plenum of the Eleventh Party Congress and then elaborated in the succeeding three years, virtually every aspect of rural organization was transformed. Prices for agricultural goods were raised, villagers got to plant more high-income commercial crops, farmland was partitioned out to families, some peasants left field agriculture entirely, and starting in 1984 party control over communes began to be replaced by economic managerial control over townships.

From this list, it is tempting to focus on the move toward family farming, and to conclude that China's experience with agriculture once again shows how the private family farm is superior to larger, collective units. It is the argument of this volume that this is too simple a view—that many of the lingering problems in Chinese agriculture in the middle-1970s had to do not with the micro-incentives of family versus collective farming but with more macro-issues of government planning and administrative strategies concerning agricultural investment, pricing, loans, and the like. This volume thus begins with these issues, analyzed by economists who examine statistics based on more than a single village. This order of presentation also coincides with how these changes were introduced in China—the macro-level planning and administrative strategies were emphasized in 1978 while the most dramatic micro-level incentive changes began after 1980.

## Planning and Administrative Strategies

The question of proper macro-incentives involves the long-debated issue of the proper role of government in economic growth. The Chinese experience over the last three decades provides examples of both constructive and destructive government intervention.

The current literature on proper modes of government intervention remains mixed. Some authors call for an active government role in building rural infrastructure, including research stations, extension services, water control works, marketing, and other support services that benefit more than any one individual farmer. Some suggest that the state can and will play a central role in late-developing societies—the challenge is only to build a strong bureaucracy free of corruption and committed to serving national as opposed to narrow personal, local, and class interests. Statistically, it has been shown that societies with strong administrative structures in the countryside have tended to have more rapid agricultural growth over the last couple of decades.[5]

Yet other authors continue to identify interventionist governments as a major source of economic distortions in rural development. To these authors, it is vain to speak of making a bureaucracy serve national interests when the urban middle classes have many more political resources than peasants. In this situation, active state intervention in the market makes it likely that urban interests in things such as industrial investment and cheap grain will be served while cultivator interests in things such as agricultural research and adequate grain prices will be ignored. Even when not serving particular class interests, bureaucracies cause problems of their own, being slower than the market to respond to new situations and serving as much to protect individual bureaucratic careers as larger social purposes. These problems are perceived as particularly acute in socialist regimes, and Soviet collective farms are often taken as an ideal-typical example of how agriculture can be stifled by the drive to fund industrialization by draining agriculture and by other sorts of improper bureaucratic interventions.[6]

It seemed for a time that China had avoided many of the latter types of problems. Based on a peasant revolution and with much talk about serving the peasants, it appeared that collective agriculture was not being used to drain resources from the countryside into cities or urban industry. And, except in the Great Leap Forward period, the state appeared to avoid heavy-handed bureaucratic management of agriculture in favor of more village autonomy, which allowed small peasant communities to make their own decisions about which crops and cultivation methods suited their area. Thus, China seemed to enjoy the benefits of many bureaucratic services from a strong central state without many of the disadvantages.

In hindsight, it appears that the peasant base of the Chinese revolution made less difference in the management of agriculture than we once thought. As Lardy explains in this volume, many aspects of the Soviet model were replicated in China in the three decades following the revolution. Even while the tax burden was lighter and the differential in prices between urban and rural goods was not so severe as in the early years of collective agriculture in the Soviet Union, prices were still slanted sufficiently in favor of the urban sector that there was a net drain out of agriculture. The state invested little in agriculture in return, even while increasingly large subsidies were being given to underwrite the food, housing, and other needs of the urban dwellers. Though committed to rural-urban equality in name, China encountered the problem of urban bias common to many developing societies and to the European socialist societies of earlier years.

Besides maintaining price and subsidy policies that favored cities, the state bureaucracy intervened rather directly in agricultural planning in a way that further lowered peasant incomes. By insisting on a grain-first, self-sufficiency policy guaranteeing that urbanites were fed and the grain-poor rural areas were not a drain on the state, villagers were prevented from going into profitable commercial corps. In a not-so-successful attempt to grow ever more grain, triple-cropping and other schemes were forced on farmers, driving up production costs as fast or faster than receipts. Though there was talk of favoring rural small-scale industry, the central control of raw material helped keep these industries at minimal levels so that they could not contribute all that much to rural income (see table 1.1, panel D). The nature of these pre-1978 problems is elaborated in chapters by Lardy and by some of the authors conducting field studies.

Other authors discuss the potentially positive role of state bureaucracy in rural development. Marc Blecher describes the government institutions serving agriculture in a county with a particularly elaborate rural infrastructure. Tom Wiens provides a systematic comparison of northern Chinese villages that have and have not succeeded as a result of government efforts. Examining the structure of government-set prices for agricultural commodities in the post-1978 period, Wiens concludes that absolute price levels are now sufficiently high to encourage production and that relative prices among different crops are now appropriate for encouraging an efficient allocation of resources. What is more problematic is whether villagers will continue to have the freedom to respond to appropriate price signals and whether the government will provide sufficient support services for agriculture. Research and extension services, facilities for maintaining seed purity, the mix of available fertilizers, the varieties of pesticides, and credit facilities remain less than ideal. The last factor is particularly important. The major factor that separates successful from unsuccessful villages in Wiens' sample is whether they have received government financial assistance in the past. Without that assistance, in the form of loans or other instruments, villages are unable to make the irrigation and other improvements that would promote their development.

In a study of another North China area, Steve Butler gives a more pessimistic account of the probable long-term successes of the immediate post-1978 reforms. Incomes had improved as prices rose, as villagers were allowed to abandon unprofitable methods, and as more farmers went into sideline activities such as making bricks or noodles. But many limitations on the ability to respond to new price signals remained. Village accounting methods remained so rudimentary that it was difficult to determine the costs for producing each crop. Village leaders were habituated to accepting orders from above passively, and agriculture was perceived as a dead-end job that attracted little leadership talent. Village leaders did not understand the new technology. The new contract system for agriculture used in Butler's area actually increased the number of targets from above that local villages had to meet. And growing grain and some other agricultural products was still perceived as no better than a break-even endeavor done only to meet state-imposed obligations and personal subsistence needs. The initial effects of the post-1978 reforms were quite positive, then, but the long-term prospects for positive change under these reforms remained problematic—particularly because it was still unclear whether local villages could operate effectively in a quasi-market environment.

After 1980, however, we were not to see whether collective units could operate successfully in that new environment. The government chose instead to convert the countryside to a kind of family farming. We can only guess at the many factors that went into this decision. Part of the answer is that national leaders who opposed collective farms in the 1950s had since returned to power and were now in a position to implement their views. More significantly, national leaders may have begun to feel that they could not afford to transform agriculture through expensive price and financial instruments alone. To do so may have been perceived as threatening both long-term industrial modernization and the implicit promises to urban residents to keep the prices of basic food comodities low. A kind of urban bias, thus, continued. Lardy notes the resistance to raising rural prices further, and Wiens notes the insufficiency of loan funds available to farmers. To move toward quasi-family farming, then, may have been seen as a cheap alternative, providing a quick fix for many of the problems afflicting agriculture.[7]

Just as importantly, the Butler account suggests, the in-between response of loosening central control and relying more on prices may have had problems. Many local leaders, it would seem, would not allow villagers to make decisions completely on their own. Untrained in responding to the market, concerned about maintaining control, perhaps ideologically predisposed to favoring bureaucratic means of administration, and fearful of what political reversals might occur in the next campaign, many local leaders continued to resist efforts to provide villagers more autonomy—or so many stories in the press suggested. It may have been, then, frustration with local leaders that led to a sudden post-1981 swing to more family farming.

The effort to sidestep old local leaders is also seen in the new administrative reforms that began to be implemented more widely at the start of 1984. Increasingly the old commune organization led by party specialists was replaced by a new township organization led by economic specialists. In theory, then, technicians were to replace politicians.

Regardless of the exact intent of national leaders, the chapters in part 1 of this volume suggest that the proper level of government intervention remains a critical issue. In some respects, through an infrastructure of roads, telephones, broadcast networks, irrigation works, health services, and education, the Chinese government has provided an essential basis for further economic progress. Recent changes in price and investment policies as well as the greater autonomy for villages in planting move in an even more productive direction. It is questionable whether these changes have been sufficient for long-run growth.

### Incentive Systems

The second part of the volume examines China's grappling with different incentive systems within a single village. Again, the conclusion is not simply that family farming is better than collective farming but that "it depends." The authors in this section examine on what it depends.

Literature on other societies provides some clue to the conditions under which collective incentives might be effective. Part of this literature notes that usually there are few advantages to scale in agriculture—a small farm is just as, and in some ways more, efficient as a small farm. With the pooling of land in a large, collective farm, some borders between fields may come into cultivation for the first time. Fields may be more thoroughly leveled and made amenable to machine cultivation. Some hidden labor with little or no land to work on may be more fully utilized. Yet these advantages are often offset by the disadvantages associated with trying to provide proper incentives for hard labor and trying to manage large production units.[8]

Still others note that there may be a basis for collective agriculture in some situations, particularly when peasants desire the security of collective sharing. In a "moral economy" approach to village life, James Scott suggests that many villagers are less concerned with individual-profit maximization than with protecting themselves against starvation in times of adversity. This protection is provided by morally enveloped village communities in which the wealthy are obligated to provide welfare support so as to keep fellow villagers above a subsistence minimum during years of famine and other difficulties. To the extent that these sorts of moral obligation already existed in villages, it should be relatively easier to find a base on which to build new collectivities.[9]

In an alternate "political economy" approach, Samuel Popkin suggests that the building of strong village ties is much more problematic. Most peasants, he argues, are not oblivious to the possible benefits of collective action, but their collective involvement is predicated not on moral obligations but on rational calculations that this involvement will benefit their individual family. They become convinced of the rewards of collective action only when they believe that others will cooperate fully in the collective endeavor, not shirking work or taking undue rewards. They are more easily convinced of this when the collective units are small, allowing for easier metering of each person's contribution and just reward. People can also become convinced when there is a self-sacrificing leader who becomes a model for emulation and who convinces the potential participants that rewards will be distributed fairly. Additionally, the provision of extra payments that can only be earned by joining the collective may help allay some of the participants' fears that they will not be justly rewarded.[10]

Louis Putterman extends this analysis with an examination of voluntary participation in collective farms in Tanzania. Much as in the earlier analysis, Tanzanian peasants were more likely to participate in collective units if these were small, permitting easy measurement of labor participation; if payment was more by labor than by need; if a strong, trusted leader handled labor rewards; if the collective sector was more productive than the private sector; and if the collective was dedicated not to subsistence farming but to more complex production such as sugar cane, fruits, and cottage industry. Other conditions such as ethnic homogeneity and lack of contact with external markets had no impact on the likelihood of joining the collective. In other words, the Tanzanian peasants behaved more like Popkin's rational actors than Scott's moral peasants. They joined collective farms not out of a quest for a communitarian past but from a calculated assessment that collectives were in their self-interest.[11]

It once seemed that China had heeded the advice of both the "moral peasant" and "rational peasant" schools. In the early 1960s, after the failures of the Great Leap Forward, the basic units of collective farming and income sharing—the teams—remained relatively small. There had been some consolidation in the late 1960s and early 1970s, but the average number of households per team remained at only about thirty-four. In these teams, families were paid not equally according to need but unequally according to how much work their members had done for the collective, as measured by daily work points. And there was a small private sector, allowing farm families to earn a little extra cash income on the side, independent of the collective. Thus, much as in the "rational peasant" model, family as well as collective interests were allowed for, and the major sources of poor work controlled. In addition, much as in the "moral economy" model, many of the families in each team were close kinsmen, and the rest were long-time village neighbors, with social ties that would seem to ease the job of mutual social control of potential laggards.

Despite these seeming advantages, the Chinese government now claims that this system was not working at all—that rewards were insufficiently linked to output, and that laziness and inefficiency were rampant. Steve Butler's chapter in this volume tends to support the "rational peasant" interpretation of collective farm participation. The peasants in his village may have wanted security most of all when they joined the collectives in the 1950s, but in the 1970s they were beyond this stage. They were more concerned with increasing incomes in an open economy. To the extent that they cooperated with the collective it was because of strong local leaders whom they could trust, and to an extent because of administrative pressures from the bureaucracy above the village.

To attend better to the "rational interests" of peasants, restrictions on private plots and the free peasant market were liberalized. The size of production teams—the group of neighbors producing and sharing income—was reduced somewhat. By 1981, 38 percent of all peasant income was from private farming, and another 10 percent was from off-farm earnings, bringing total noncollective income to almost half of all rural income. The size of the average team had been reduced to thirty households from a high of thirty-four just a few years before.

These changes were but window dressing, however, compared to the changes that were to take place afterward. On the eve of the new reforms in incentives, most villages were using some variant of time or task rates to divide the year's income after the fall harvest. Prior to the harvest, the village kept a record of how much each person worked each day, assigning either a fixed amount of points according to a time rate or a variable amount according to how difficult a task had been accomplished. The points a family earned during the year were then used in dividing the distributable income of the village. The particular system used in any one village varied over time, depending on the political climate and the specific labor and land situation of that village. Jonathan Unger gives an example of the evolution of these systems in a village that tended toward a politically more moderate course, not putting much emphasis on the more egalitarian distribution systems except for brief periods of time. Steve Butler reports on a village that remained more egalitarian, employing the Dazhai system described by Unger until an unusually late 1978 date.

Government leaders decided that these earlier reward systems were inadequate on two grounds. One, these earlier systems divided farm work and income among too large a group. Even in a team of only about thirty households with perhaps sixty laborers, the results of a single individual's effort tended to be smothered so that neither extra output nor extra income from harder work was obvious. To correct this problem, responsibility for cultivation and income needed to be assigned to an even smaller unit—be that unit a small group of eight to ten individuals, much like the early 1950s mutual aid teams; an individual; or a family. Second, government leaders decided that regardless of the size of the group, rewards needed to be more closely linked to output. Thus, they urged that quotas for expected output be established and that farmers begin to sign contracts guaranteeing this output. If the guaranteed output were exceeded, the contracting farmers would keep the excess for their own use. If less than the guaranteed output were produced, then the contracting farmers would be penalized.

The systems created to cope with the perceived problems in agriculture have thus varied on two major dimensions—whether reward was linked to output and the size of the unit to which production responsibility and income have been entrusted. Table 1.2 arranges villages (production teams) according to how they vary on these dimensions. Villages have moved rapidly from the top to the bottom of this table. Most villages began with some variant of time or task rates paid to all families in the village and with no special rewards or penalties for meeting production targets. But by 1981 very few were paying a simple time rate and not many more were offering even task rates. A few offered only production contracts for specialized tasks such as making bricks, driving a tractor, cultivating an orchard, or overseeing a fish pond. Often offered through competitive bidding among village members, these sorts of contracts for specialized tasks tend to be used even when the grain fields are dealt with by one of the methods below.[12] Another intermediate approach was to contract the production of grain

Table 1.2

## Labor Reward Systems by Date

| | % of teams as of | | | |
| --- | --- | --- | --- | --- |
| | Early 1981 | October 1981 | August 1982 | Summer 1983 |
| A. Time rates and other egalitarian systems | (3.4%) | (2.2%) | | (1.7%) |
| B. Task rates | 27.2 | 16.5 | | |
| C. Rewards linked to yields | | | | |
|   1. For certain specialized tasks only | 7.7 | 5.9 | | |
|   2. Contract for output with: | | | | |
|     a. Work-group (lian chan dao zu) | 13.7 | 10.8 | | |
|     b. Individual (lian chan dao lao) | 14.4 | 15.8 | | |
|     c. Household (bao chan dao hu) | 16.9 | 7.1 | 74.0% | 95.0% |
| D. Total household responsibility (bao gan dao hu) | 11.3 | 38.0 | | |
| E. Other systems, including private farming | (5.4) | (3.7) | | |
| Total | 100.0% | 100.0% | | |

*Source: Wenchai bao*, 15 December 1981, p. 1. *Zhongguo jingji nianjian*, 1982 (Beijing), pp. V:11-12. *People's Daily*, 22 August 1982, p. 1. *Beijing Review*, no. 34 (1983), p. 7.

*Note*: Figures in parentheses are inferred from totals stated in original source.

and other staples to small groups or even to individuals for periods of one to three years. This type of contracting did not reduce the significance of the team. Indeed, as Steve Butler describes for his village, which was using this type of intermediate system in 1980, the administrative role of the team and brigade actually increased. Much of the planning for planting, irrigation, fertilization, fertilizer application, and other tasks remained under team control. And all sorts of production targets had to be set and accurate records kept for each group or individual.

With household production contracts, the next type, many of the detailed planning decisions were passed down to the family. The family thus gained more control over what it planted and what its income would be at the end of the year. The team continued to allocate work points, but only in a block as contracted at the start of the year, and to handle most relations with the state above, including purchasing fertil-izer and other inputs as well as delivering tax grain at the end of the year.

Under the next, increasingly widespread, total household responsi-bility system, the team forsook most remaining management functions. Records of workpoint earnings were no longer kept. State tax obligations, state grain-purchase quotas, and the other expenses and profits that go with running a farm devolved on the individual house-hold. A small amount was paid to the collective as an administrative fee for the few leaders who still served and for the welfare fund that helped support local medical, education, and public aid expenses. But because most of the administrative apparatus that kept accounts and handled relations with superior bodies could now disappear, and because medi-cal and education expenses were more tightly controlled, these fees were much smaller than in the past. Because of greater administrative simplicity, the immediately previous household production contract system has tended to devolve into this system.

In this system, a family is assigned fields along with farm animals and most equipment for a period initially of two to three years and now increasingly up to fifteen years. The team continues to hold pro forma title to the land and can rotate it in fifteen years. The chances of becoming a rich peasant from grain production alone are limited, since one cannot continually add to one's land holdings. Quotas for produc-tion of grain are typically imposed, and the state controls fertilizer, machines, and most other inputs. Thus, in some respects, current reforms stop short of full family farming. The present system resem-bles more one of tenant farming, with the state and collective being benevolent landlords who distribute land equally. But in many other respects the reformed farming system resembles the system of family farming that existed immediately after land reform in the early 1950s. By 1983, 95 percent of all villages had adopted one of these last two forms of quasi-family farming (see table 1.2).

In the two or more years when family farming was beginning to sweep the country it was possible to get some idea of where family and collective farming were most advantageous. In his chapter in this volume, David Zweig neatly captures the appropriate contrasts among three different villages. Among his villages poverty influenced how quickly collective agriculture was abandoned. Much as described in the press, peasants in poor areas appeared to place little faith in the collective and more in the efforts of their own family. The collective provided little income. And with grain the usual crop in these areas, there were few economies of scale. Farmers could grow grain on small fields by themselves just about as well while avoiding many of the management and incentive problems of large collectives. It would appear that in many of these areas, the collective rapidly became little more than a shell that came alive only once every few years when it was time to redistribute land among families. Much as in the Tanzanian experiment with voluntary collective farming, villages with only grain production generated little enthusiasm for collective activities.

In areas with a more diversified economy, Zweig implies, the story was quite different for a while. The brigade or team with its own industries, fish ponds, orchards, and other income-earning units had some means for retaining the loyalty of its members. Families may have preferred that their best laborers join these kinds of collective under-takings rather than work in family fields. And the collective with many supplementary-income-earning units could offer extra health, educa-tion, and welfare benefits that would continue to attract the attention of its members. Again, the pattern appears to mirror that found in Tanza-nia. Peasants calculate rationally when the collective is in their best interests and act accordingly.

While David Zweig describes some rural units that temporarily resisted the rapid transition to quasi-family farming, Victor Nee de-scribes just the opposite. His village, distant from urban markets, with few products other than grain, and with an income that was slightly below the provincial average eagerly embraced the return to a greater emphasis on family farming. He finds the sources for this return in the long-standing inability of the collective to provide rewards that a family could not provide on its own. Despite real advances in some areas, the family continued to be responsible for old age support and many other aspects of its own welfare. With the economy still depending primarily on grain production on fields scattered in mountain passes, there were few economies of scale. Workpoints were too narrowly constricted in range to reward those who felt themselves good workers. And people felt that the collective caused a stagnation in living standards. The collective spirit that had existed prior to the Great Leap Forward was seriously eroded.

In general, then, the case studies in this section tend to support the "rational peasant" model of collective farm behavior. When the collective supports a complex economy that provides many additional benefits that can not be achieved by individuals acting alone, peasants are eager to join. But in most areas, where the economy remains one of simple grain production, there are few economies of scale or other sorts of endeavors that would entice peasants into collective activities. Most of those families with ample labor power would prefer to work on their own without others organizing their work or sharing their income with neighbors.

Since the time of these studies, virtually all peasants have been forced to disburse fields and equipment regardless of their individual preferences, but for a short period we were allowed to see what direction their preferences pointed. From this period we see that the issue is not one simply of collective versus family farming but one of the conditions under which one versus another system will be favorable to the rational interests of peasants. The chapters in this section of our volume more clearly specify the nature of these conditions.

### New Patterns of Equality and Inequality

Recent economic improvements in Chinese agriculture have been dramatic. Both production and income have increased rapidly, with a much wider variety of foods being available and many additional employment opportunities in local industry and other kinds of sidelines being provided (see table 1.1, last column). Much more problematic, however, is the question of these results for many of China's social goals. One of these goals has been income equality and provision of an adequate level of income, education, and health care for everyone.

The conclusion one draws about pre-1978 rural inequality varies dramatically depending on whether one chooses to emphasize local inequality within a single village or production team, regional inequality among different villages, or urban-rural inequality. As Mark Selden discusses in this volume, local inequality was sharply reduced in the 1950s, first by land reform and then by collectivization. This increasing equality applied not only to current income but also to guarantees of basic consumption, health, and education that collective units came to provide. Most teams allocated part of their grain on the basis of need so that a family could draw grain regardless of whether it could afford to pay for this grain. And, as we have already suggested, education and health care came to be widely available for everyone in all parts of rural China. As Victor Nee explains, education and health care were not completely free, but they were often subsidized from collective sources and poor families could often get tuition remissions and help with meeting medical emergencies.

It is the latter sorts of services that are jeopardized by the new quasi-family farming arrangements in the countryside. Access to basic grain rations for those in need is not quite so automatic, and education and health facilities are more difficult to fund. When workpoint systems are abandoned in the total household responsibility system, there is no established way to pay teachers. The state has stepped in to help subsidize the salaries of some local teachers, but whether this will be sufficient to keep rural education at former levels remains to be seen.[13] Collective medical-insurance systems have also suffered. By 1981, only 58 percent of all brigades had a medical-insurance program, down from a high of about 80 percent in earlier years.[14]

. Family farming has been problematic not only because a few village families at the bottom have less security than in the past but also because there has been jealousy against those few families who have profited enormously by the new economic freedoms. At times fellow villagers have forced these families to give up their new trucks, tractors, and other signs of wealth, and these have been returned only when the outside government stepped in.

However, these problematic changes in local village equality must be set over against other changes in regional inequality. As Selden, Lardy, and Diamond note in this volume, the differences in income between villages remained severe through the 1970s, and may have even increased since the 1950s. Some of the intervillage differences were the result of inequalities in land holding and urban proximity that predated collectivization. Others were the result of administrative intervention in later years. Norma Diamond describes a village consigned to the periphery of an administrative region and therefore neglected in the allocation of many resources and in planting allocations for more profitable crops. Both Lardy and Diamond describe how the obligation to grow grain even in villages better suited to other crops caused stagnation and even poverty in some villages. The relaxation of the obligations on growing grain allowed some poor villages to recover rapidly after 1978 as they moved into raising animals and growing cotton, tobacco, and other commercial crops. These changes help explain how so many peasants escaped the poverty level of 100 yuan per capita income after 1978.[15] Thus, when local inequality is combined with regional inequality, the pattern of inequality throughout the Chinese countryside may be no more severe today than in the past when the largest gaps in income were among regions rather than within a single village.

The gap between the city and countryside is also problematic. Recently released statistics in China show that the gap in consumption between city and countryside remained at about 2.6:1 throughout most of the last three decades.[16] The radical talk about favoring peasants over urbanites had no impact on that gap at all. Marc Blecher dissects the nature of this gap at the county level. He shows that by building in part on the old structure of small cities and market towns spread throughout the Chinese countryside even before the Communist takeover, the government was well positioned to bring new service organizations into close proximity to the average villager. Many of the marketing coops, savings banks, and other institutions he describes are located in market towns in easy reach of every villager and in places that the villager was accustomed to coming to anyway. Also, at the county level, taxes and expenditures appear to be approximately balanced between local urban and rural interests, with the county government sometimes helping on major rural water conservancy projects.

Much more troublesome is the line the government draws between urban and rural households. As Marc Blecher and several other authors explain, residents inherit an agricultural or nonagricultural household registration at birth. To be born with an agricultural registration severely limits one's life chances. To limit the growth of cities and urban infrastructure expenses, the barriers to changing one's registration from agricultural (rural) to nonagricultural (urban) are many. Few rural residents are allowed to move to the city except as temporary contract laborers in industry, where they often earn lower wages and fewer fringe benefits than regular urban laborers.[17] Blecher argues, nevertheless, that at the county level, the gaps between regular and temporary contract laborers are not all that great, and that the contract system has certain benefits for the countryside as well as cities. By insuring a regular flow of cash remittances to spouse, children, and other kinsmen left in the village, the contract system helps increase rural living standards above what they might otherwise be. These and other possible advantages help make the strict line between urban and rural residents a little less harsh in its consequences than it might otherwise seem, Blecher argues.

In total, then, China is still struggling with the proper balance between urban and rural rewards. This volume begins with Nick Lardy's discussion of how China has followed the early Soviet model of using the countryside to support urban interests for almost three decades and of how current policies help alleviate, but do not completely solve, this imbalance. The volume concludes with Marc Blecher's chapter suggesting that at the county level, the problem of urban and rural imbalance is not quite so severe and that some of the harsh measures taken to keep peasants out of cities on a permanent basis may have some hidden benefits for both sectors.

## Conclusion

What are the lessons, then, that China's experience with agricultural development suggests? In part, this experience suggests that there is no easy answer to the many goals that might be pursued in agriculture. Some of the goals desired in China, and in many other parts of the world as well, are clearly in contradiction. The pursuit of a highly egalitarian rural society with ample education, medical care, and other kinds of support for all may well conflict with rapid agricultural growth. China pursued its goals of health care, basic education, subsistence floors, and income equality in part on the basis of local self-help efforts. The central state apparatus invested little more in these sorts of activities than did most other developing societies over the last three decades. In this manner, it could still have funds for high levels of investment in industry even while producing more income equality and support services within villages as neighbor shared with neighbor through collective team and brigade structures. Some major gaps among villages and between city and countryside were allowed to remain, but to an unusual extent rapid industrial growth was pursued even while some degree of rural equality and social services were provided.

It is this mixed system that is now in jeopardy. With much weaker collective structures in the countryside, it is unclear that most villages will be able to provide the locally funded schoolhouses, teachers, barefoot paramedics, grain allowances, and other kinds of relief that they did in the past. The central government has as yet indicated no willingness to assume these additional expenditures, even though the problems of declining rural support for social services has been mentioned in the press. To do so would jeopardize the "four modernizations" campaign for the urban economy.[18]

Without such support, some families with fewer male members or with fewer skills may begin to complain that they have been left behind in the rush to introduce market mechanisms and quasi-family farming. These complaints would echo the complaints Unger reports for his village in the early 1960s when a similar form of family farming was introduced. The political consequences, though perhaps less shrill now with more of the original revolutionary generation out of the way and with peasants generally at a higher standard of living, could be much the same, with future flip-flops in policy as the government tried to hit upon yet another ideal solution.

Those flip-flops will be encouraged by some other, less than benign consequences of current policies.[19] One is the tendency for the rural birth rate to rise as control by local administrators is weakened and as families begin to calculate that they can do better on their assigned fields with an extra son. Not controlling workpoints or many other resources, there is little that the collective can do against families who adopt this kind of rationale. Even when collective units can pressure families by withholding fertilizer, cooking oil, cotton, sugar, and other goods not produced locally, the consequences can be unpleasant. Faced with these sorts of choices, a certain proportion of families, one hopes small, has begun to kill their infant daughters. The spector of a new wave of infanticide, already discussed in public by central leaders, cannot help but raise major warnings about how current policies return one to a feudal past. This gives orthodox socialists a ready weapon for arguing that a brake must be applied on the drift of current policies.[20]

Another negative social consequence is for some parents to take their children out of school in order to help tend the family's new fields. As a consequence, enrollment has declined during the last two years in some rural schools. The drop is particularly severe in upper and lower middle school, but some primary school children as well are now said to be dropping out in order to help increase current family income.

Certain economic goals are also threatened. With the relaxation of administrative control in the countryside, the willingness to allow peasants to engage in activities other than growing grain, and the relaxation of urban food supplies, it has become far easier for peasants to move to nearby cities and towns even without a proper work or residence permit. The sharp dividing line between urban and rural that Blecher describes is going to be far more difficult to enforce in the future. With its commitment to maintain urban living standards, the state may have to spend even more funds on urban schools, roads, houses, parks, and cultural facilities in order to meet the needs of new urban residents, both legal and illegal. The proposed solution is to promote the growth of small towns and cities, but whether this will be sufficient to stem the incipient tide remains to be seen.

Another problem is that in some areas the emphasis on more freedom of action has meant that families have begun to denude hillsides in pursuit of grazing lands for their private animals, for fuel, and for timber for the new housing boom in the countryside. The result is increased erosion, which threatens long-term productivity and the control of silting and floods.

There are also complaints that the area sown to grain has dropped to early 1950s' levels as peasants have fled grain production for better paying commercial crops such as cotton, tobacco, and soybeans. Tighter controls must be imposed to insure that enough grain is grown, it is said.[21] One suspects that it is also more difficult to pool resources for large public works projects such as the irrigation and flood control works described by Blecher in this volume. Such projects may have been overdone in the past, with insufficient reckoning as to whether they would pay off. But whether they can be mounted now, even with careful reckoning, is not clear.

Also, it has not been shown in detail that quasi-family farming is really all that necessary for many of the very substantial economic gains that have been made since 1978. As the papers in the first part of this volume suggest, many of the problems in Chinese agriculture have had to do not with insufficient incentives internal to a village but with the constraints imposed on the village from outside. Once many of these "macro" issues were solved, production and incomes increased rapidly after 1978. The most significant of these increases were achieved even before the shift to quasi-family farming in 1981–82. Increased prices, less external interference, and a freedom to raise more profitable crops suited to one's area account for much of the spurt in production since 1978. As the Polish experiment with family farming shows us, micro-incentives without proper attention to the macro-incentives imposed from outside the village provide no sure path to growth in agriculture.

The studies in this volume and published studies from China provide as yet no firm answer as to the relative contribution of changes in micro- and macro-structures in agriculture. But as China's leaders begin to reassess the total range of consequences of the changes of the last five years, and rural progress over the last three decades, they may well conclude with us that there are no easy answers on how to pursue agricultural development in such a way that all groups and most social goals are well attended to.

*Postscript on Sample Villages*

The villages represented in this volume are hardly a representative sample of all villages in China. As indicated in table 1.3, the villages in our case studies tend to have higher than average income, higher grain yields, and a rather high proportion of all income from collective "side

Table 1.3

## National, Province, and Sample Village Characteristics

| Locale | Investigator | Year | Yuan per capita* | Grain yields (ton/ha.) | Collective side activities: % income | % labor | Private income % | Model village | Major crops | Urban proximity (kms.) |
|---|---|---|---|---|---|---|---|---|---|---|
| National figures | | 1978 | 74 | | | | 26 | | | |
| | | 1979 | 83 | | 15 | | | | | |
| | | 1980 | 86 | | | | | | | |
| Hopei province | | | 84 | 3.5 | | | | | | |
| Shulu county | Blecher | 1980 | 88 | 6.0 | 33† | | | Partly | Grain, cotton | Distant (60) |
| Dahe commune | Butler | 1979 | 126 | 9.0 | 52† | | | No | Grain, cotton | Medium (20) |
| Wugong brigade | Selden | 1978 | 188 | 8.0 | 43 | 15-20 | 6 | Yes | Grain, cotton | Distant (120) |
| Shandong province | | | 102 | 4.8 | | | | | | |
| Team A | Wiens | 1980 | 28-70 | | 0-1 | 0-8 | | No | Grain | Distant |
| Team B | Wiens | 1980 | 117-130 | | 0-1 | 0-9 | | No | Grain, cotton | Distant |
| Taitou brigade | Diamond | 1978 | 132 | 3.1 | 38 | 12 | | No | Grain, sweet potatoes | Medium |

| | | | | | | | | | |
|---|---|---|---|---|---|---|---|---|---|
| Jiangsu province | | 1980 | 102 | | | | | | |
| September 1st brigade | Zweig | 1980 | 120 | 7.3 | 2 | 17 | No | Grain, cotton | Distant (50) |
| Qingxiu brigade | Zweig | 1980 | 230 | 4.8 | 50 | 33 > | Yes | Grain | Medium (30) |
| Mu shuyuan brigade | Zweig | 1980 | 320 | | 63 | 33 > | Yes | Vegetables | Suburban |
| Fujian province | | 1980 | 72 | | | | | | |
| Yangbei brigade | Nee | 1979 | 68 | | | | No | Grain | Distant |
| Guangdong province | | 1980 | 105 | | | | | | |
| Chen brigade | Unger | 1973 | 90 | 5.2 | 17 | 17 | No | Grain | Distant |
| | | 1981 | 400 | 5.2 | | 2 | No | Vegetables, grain | Medium‡ |

*Sources:* *Zhongguo baike nianjian*, 1981 (Beijing), *Zhongguo jingji nianjian*, 1981 (Beijing), p. VI-10; Individual field studies.

*Notes:* *Collectively distributed income alone, in most instances, thereby excluding income earned in the private sector.

†These two figures include income earned in commune industry, thereby inflating these estimates compared to the rest in this column.

‡By 1981, Chen village had received approval to sell vegetables in Hong Kong, hence the change in distance to a city.

activities" such as making bricks or raising ducks. Some of them are model villages held up by national and regional governments for others to emulate—and which have often received special government assistance. Yet while many of the villages in this volume are specially blessed, others are much more nearly average for their region or the nation as a whole. For example, the villages studied by Tom Wiens are near average and even below average for the region in which they are located. The villages studied by Victor Nee and by Jonathan Unger are probably no better than the average for their regions. This variation among regions, then, provides an additional opportunity to study the sources of policy variation (as in David Zweig's study) and of prosperity (as in Tom Wiens's study). These variations in income, yields, collective sidelines, and the variety of crops grown help explain why some villages remain very leery of collective activities (as in Victor Nee's village) or reluctant to abandon collective farming (as in two of David Zweig's villages). Overall, then, the villages represented provide a good basis for examining the rural changes taking place in China since 1978.

## Notes

1. *Beijing Review* 39 (1983):4. Other statistics show that this shift was more than just an artifact of changing prices.
2. See Martin K. Whyte and William L. Parish, *Urban Life in Contemporary China* (Chicago: University of Chicago Press, 1984), ch. 4.
3. Ibid.
4. *Zhongguo nongye nianjian, 1980* (Beijing: Agricultural Press, 1981), p. 41.
5. On the need for a strong bureaucracy, see Gunnar Myrdal, *The Challenge of World Poverty* (New York: Vintage, 1970). On strong administrative structures and agricultural growth, see Norman T. Uphoff and Milton J. Esman, *Local Organization for Rural Development* (Ithaca, N.Y.: Rural Development Committee, Center for International Studies, 1974).
6. On the general problem, see Theodore Schultz, ed., *Distortions of Agricultural Incentives* (Bloomington: Indiana University Press, 1978). On urban bias, see Michael Lipton, *Why Poor People Stay Poor: Urban Bias in World Development* (Cambridge: Harvard University Press, 1977). On problems with Soviet collective agriculture, see Michael Ellman, *Socialist Planning* (Cambridge: Cambridge University Press, 1979), ch. 4.
7. Another indication of the desire to do things cheaply was the decision to cut national agricultural investment funds from an already low 10 percent to only 6 percent of state investment in 1982. State Statistical Bureau, *Chinese Statistical Annual, 1983* (Hong Kong: Economic Reporter Press, 1983), p. 325. One explanation was, of course, that with increased prosperity, peasants could not generate investment funds on their own—but see Wiens's suggestions in this volume.
8. For example, see the discussion in Michael Ellman, *Socialist Planning*, ch. 4.
9. James C. Scott, *The Moral Economy of the Peasant* (New Haven: Yale University Press, 1976).

10. Samuel Popkin, *The Rational Peasant* (Berkeley: University of California Press, 1979).

11. Louis Putterman, "Is a Democratic Collective Agriculture Possible?" *Journal of Development Economics* 9 (1981):375–403.

12. With the increased liberalization in agriculture, many peasants abandoned grain cultivation for the kinds of specialized tasks described here. By the start of 1983, among a total agricultural labor force of 320 million almost a third were in nongrain activities—of these about 30 million were in small factories and about 70 milion were in other activities such as fish and poultry breeding, transport, and commerce. About 10 percent of all agricultural households had all their members in specialized, nongrain growing tasks. See *Beijing Review* 13 (1983):5, 22.

13. See ibid. 4 (1983):25. Rural teachers are divided between state teachers on the state payroll and local teachers on the village workpoint system. It is the latter type of teacher that now presents a problem.

14. *Jingji nianjian, 1982* (Beijing), p. v-392.

15. See note 1.

16. *Jingji nianjian, 1982*, p. vi-25.

17. In China, the urban population grew from 18 percent of the total population to only 21 percent in 1982—a minimal increase compared to other developing countries. It seems that almost one-third of the 1982 urbanites maintained an agricultural registration because they farmed nearby fields or worked only as temporary laborers in the city. The percentage reported as urban varied between 13 and 21 percent, depending on whether the agricultural part of the urban population was included. See *Beijing Review* 7 (1983):28.

18. The Soviet Union did not begin to provide pensions and other kinds of additional support for collective farmers until the late 1960s, suggesting that it takes a rather high level of economic development before states are able and willing to take on these additional burdens. See Victor George and Nick Manning, *Socialism, Social Welfare, and the Soviet Union* (London: Routledge and Kegan Paul, 1980). Lacking state support, Chinese villages themselves provide regularized pensions to no more than 1 percent of all aged residents (*Beijing Review* 27, 30 (1984):9.

19. Reflecting the second thoughts that have arisen in China since 1981, Victor Nee provides a useful list of the negative side-effects of current policies that have recently surfaced in the press. See his "The Peasant Household Economy and Decollectivization in China," paper presented at the annual meeting of the Association for Asian Studies, San Francisco, 1983.

20. There is a partial way out of this dilemma, which is for the state to assume the burden of old-age support rather than forcing it upon rural sons. But, as already suggested, this would threaten the competing goal of urban modernization. Chinese academicians have begun to point to the negative consequences of this structural problem for both birth control and female infanticide—e.g., *Shehui* (Shanghai) 3, 2 (1983); Cheng Du, "Rural Fertility in Hubei Province," *Renkou yanjiu* 5 (1982):36–38, 31.

21. *Beijing Review* 14 (1983):24.

# PART ONE

## Planning and Administrative Strategies

# 2

# State Intervention and Peasant Opportunities

# Nicholas R. Lardy

Many Western studies of contemporary China presume a close relationship and an interdependence between the Chinese Communist Party and the Chinese peasantry. The party's early effort in the 1920s to mobilize support in urban China is regarded both by the party and by Western observers to have failed.[1] An important school of thought attributes the rise to power of the Chinese Communist Party in 1949 largely to its shift after 1927 to a rural-based strategy and its ability to respond to and represent the revolutionary interests of the majority of Chinese peasants.[2] The success of party policies in the first five years of the People's Republic, particularly of the land reform program and the formation of small-scale farming cooperatives, generally has been thought to rest on the ability of these programs to meet the aspirations of the peasant majority, particularly the so-called low and middle peasantry. Charles Roll's work, for example, shows that land reform increased substantially the incomes of the nonrich peasant majority, largely by reducing the incomes of former rich peasants and landlords.[3] The growth of farm output and peasant income and consumption during this period is supported by both Chinese data and Western studies.[4] The contrast with earlier Soviet programs, which depressed farm production, income, and consumption and led to widespread loss of life, is usually attributed to the intimate familiarity of the Chinese Communist Party with rural conditions and peasant interests and juxtaposed against the Soviet party, which had risen to power on the basis of a largely urban-based revolutionary movement and was insufficiently sensitive to peasant aspirations.[5] Some have gone so far as to assert that the evolution of rural institutions and peasant-state relations, for at least the first two decades of Communist rule, can be seen as furthering the interests of the peasant majority.[6]

In this view the Great Leap Forward was an aberration attributable either to Mao Zedong or to local party cadres who were excessively zealous in their pursuit of the vision of the communistic society that was being projected by the party leadership in Beijing. Once the failure of these policies was evident, the party adopted an "agriculture first" strategy in which rural development was to be achieved not through normative appeals but through incentive policies based on peasant self-interest, combined with a recognition that state industry would have to supply farmers with significant quantities of modern industrial inputs such as chemical fertilizers, pesticides, and certain types of farm machinery. The degree of socialism in the countryside was reduced as increased authority was devolved from communes to production teams, relatively small rural units with about twenty-five households, equivalent to the lower-stage cooperatives that were the predominant form of organization prior to collectivization in 1955–56.

The relative stability of rural institutional arrangements throughout most of the 1960s and 1970s underlay the widespread assumption that production units in rural China, most commonly teams, allocated resources efficiently, much as would farm households in a market economy. The agricultural sector was presumed to produce the maximum potential output given the prices set by the state for farm products and for current inputs, the constraint imposed by the available supply of land, and farmers' labor-leisure preferences. In the words of one writer, "there are no obvious and gross inefficiencies in Chinese farming."[7]

The view advanced here is that, on balance, since 1949 state policies have been inimical to peasant interests. This judgment rests not on an evaluation of policies pursued during periods of limited duration such as collectivization in 1955–56 or the Great Leap Forward, which are recognized widely to have been detrimental to peasant livelihood, but rather on a long-term tendency for the state systematically to constrain peasant income-earning opportunities and to inhibit the efficient use of resources in agriculture. The focus is not on inefficiency arising from the formation of producing units so large that the connection between individual effort and reward became tenuous, exacerbating the free-rider problem common to agricultural producer cooperatives in which production activities are dispersed and the costs of monitoring individual work effort prhibitive. These are essentially what Liebenstein refers to as a source of x-inefficiency, arising from the failure to provide satisfactory motivation for individuals.[8] Rather, the focus here is on how state price and marketing policy have affected allocative efficiency.

State interventions in the production and marketing of farm crops were initially quite modest. Following land reform in the early 1950s, over 100 million farm families cultivated their own land and the state had no direct means of controlling their production decisions. The central tenets of state policy were to increase farm output and restore the marketing system, primarily through rejuvenating private marketing activity and restoring price incentives. As early as 1950 central authorities sought to achieve the desired composition of farm output by manipulating relative crop prices. That was achieved by announcing that, for state purchases, certain minimum price ratios would prevail between individual commercial crops, such as cotton and other fiber crops, tea, tobacco, and some oilseed crops, and the major grain crop of each region.

Prices of cereal crops, however, were market determined, and state purchases were a relatively small share of agricultural output, so state intervention probably did not depress farm prices. Rural periodic markets, which had facilitated commercialization of Chinese agriculture since at least the time of the transition between the Tang and Song dynasties, flourished. Even after price stability for industrial products was restored, farm prices continued to rise, so that farmers' terms of trade and income rose. Interregional trade in agricultural commodities recovered during the early 1950s as well. These interregional commodity flows not only helped to alleviate temporary food shortfalls caused by locally adverse weather conditions, but also facilitated specialized production of noncereal crops. Resurgence of agricultural output during the early 1950s, of course, partially was a consequence of the restoration of normalcy following a sustained period of war and civil war, but state commercial and price policy on balance facilitated rather than inhibited recovery.

State intervention in farm production and marketing decisions increased drastically in the fall of 1953 when a system of compulsory deliveries of agricultural products at fixed prices was instituted.[9] Prior to 1953 most deliveries to the state, above the large amounts collected in kind as taxes, were voluntary. It appears, however, that the state feared a substantial further run-up in the price of cereals as the large-scale industrialization program of the first five-year plan got underway. The prices of grain and other agricultural products were critical because cereals were the major wage goods, and upward pressure on grain prices would give rise to the need for further increases in wages of workers in both industry and other urban employment, increasing the state wage bill and thus reducing the resources available for investment. Moreover, agricultural exports were essential to finance the capital goods imports from the Soviet Union that were to form the heart of the first plan. The early efforts of the Chinese Communists to arrange substantial grants and credits from the Soviets had been largely unsuccessful. The Soviets were willing to supply advanced capital goods and technical assistance, but they were unwilling to provide long-term credits. Rather, purchases of Soviet machinery and equipment as well as technical assistance would have to be financed primarily by exports, predominantly food and textile products.[10] Thus the rate of capital formation was even more directly linked to the level of prices the state would have to pay the peasantry for the major export goods.

In theory, a system combining forced deliveries to the state at fixed prices and a private market need not depress the average price received by peasants below the price that would prevail in a single free market with no government intervention.[11] The average price received under a mixed system is simply the weighted price received for compulsory deliveries to the state and the open market price. If the open market price rises sufficiently in response to the supply withdrawn by government procurement at fixed prices, the average price received by farmers need not fall. Indeed in the short run, when total supply is fixed, the weighted average price might even rise.

When the system of compulsory deliveries for cereals was begun in the fall of 1953, after fulfilling state quotas, peasants nominally still were allowed to sell surplus grain beyond their own requirements for consumption, seed, and animal feed. In the absence of other restrictions the introduction of compulsory sales to the state need not have reduced the average price received by producers and thus need not have undermined production incentives or reduced peasant income. Private marketing, however, was restricted substantially by two policies. First, private grain millers were no longer permitted either to purchase grain in rural markets or to sell grain products privately, but became subject to the direct administrative control of the Ministry of Food and its local purchase and sale agencies.[12] Second, traditional grain markets in both cities and market towns were converted to "state grain markets."[13] These two policies effectively limited private transactions to small-volume transactions within geographically very limited local rural markets. Trade over broad geographic regions or in larger quantities became a monopoly of the state and thus subject to state price control. The restrictive effect of those policies was reflected in a sharp decline in private grain sales, from an average of seven to eight million metric tons in the early 1950s to two to three million tons by 1954–55.[14]

Restrictions on the marketing of cotton, the single most important nonfood crop, were more severe than those for grains. While a large share of grains produced were retained for self-consumption, after compulsory deliveries of cotton were instituted in 1954, peasants engaged in cotton production were allowed to retain only limited amounts of raw cotton, initially only about one to two kilograms per capita. Even these limited quantities of cotton could not be sold on rural markets, either as raw cotton or in processed form as homespun yarn or handwoven cloth. Private sale of all of these was legally prohibited on periodic rural markets.[15] Rural handicraft cloth production, which had survived several decades of competition from both foreign and Chinese-owned modern factories, was suppressed by the price and marketing policies adopted in the mid-1950s, substantially curtailing the income-earning opportunities traditionally open to peasants in cotton-producing areas. The state simultaneously became the monopolist purchaser of raw cotton and the monopolist seller of cotton textile products. Moreover, the state set the prices of raw cotton and of finished textile products in such a way as to further tax the farm sector, which was the source of the raw materials for the textile industry and, since 85 percent of the population was rural, the major market for textile products. State pricing policy quickly established the state-run textile industry as the single most important source of government revenue.[16]

In brief, then, by effectively limiting private trade in cereals and eliminating most private trade in cotton, compulsory purchase schemes reduced the market opportunities available to peasants. A larger share of transactions was with the state at prices set primarily with a view to generating budgetary revenues to raise the rate of capital formation. Income-earning opportunities from domestic textile production by handicraft methods were curtailed since production was limited to immediate family needs.

In addition to reducing the supply of agricultural commodities to private markets by requiring peasants to sell most of their surplus to the state at fixed prices and placing restrictions on the opportunities for peasants to dispose of remaining supplies on private markets, the state in 1955 made rationing of cereals and edible vegetable oils universal in urban areas. The coupon rationing system introduced supplied quantities of cereals and edible vegetable oils that were sufficient to meet normal consumption demand. That reduced the opportunity for peasants to sell these products directly to urban consumers through state-controlled grain markets.[17]

State intervention increased during what is known as the "socialist high tide" in the winter of 1955–56 when virtually overnight the state abandoned its voluntarist approach to cooperative farming and instituted collective farms with an average size of 150 families. Simultaneously most remaining private commerce was "transformed" into state ownership and control and rural periodic markets were closed. That placed a squeeze on the remaining private activities within agriculture, in some regions substantially reducing peasant income. Fei Xiaotong, China's best-known anthropologist, after returning in April and May 1957 to a Jiangsu village he had first studied in 1936, criticized the curtailment of these income-earning opportunities, stating that it had depressed peasant income.[18]

The formation of large-scale collectives also made it possible for the state to attempt to plan agricultural output directly, much as it was doing in industry. Rather than manipulating relative crop prices to influence cropping patterns, planning, at least in theory, was to be based on explicit sown area and output targets drawn up at the national level and then disaggregated and transmitted down the administrative hierarchy to provinces, then to counties, and ultimately to the newly established collectives. That direct approach to agricultural planning, of course, was infeasible under a system composed of more than 100 million private family farms or even of the 10 million mutual aid teams and lower stage agricultural producer coops that existed in 1954 and the first part of 1955.

Although collectives had been formed so hurriedly that in some regions they were no more than paper organizations, direct planning of sown area and the imposition of certain techniques were attempted. In 1956, for example, the state forced cooperatives to purchase over a million of the infamous double-wheel, double-bladed ploughs. Although the implement earlier had been used with some success in areas of dryland farming in the north, in 1956 the plough was forced on collectives in central and southern China, where it was virtually useless in paddy cultivation. At this time also provincial- and county-level authorities dictated specific planting densities for rice, irrespective of varying local conditions. "The planning system was over centralized to the point where collectives merely mechanically carried out orders from above, especially those received from the hsien (county) authority."[19]

The adverse affect of both centralized determination of cropping patterns and production techniques and curtailed periodic rural markets led to pressure for modifications. In the summer of 1956 the state officially sanctioned the reopening of rural markets, and in September at the Eighth Party Congress there was explicit criticism of the movement toward production planning.[20] Leading party members concerned with agricultural development policy, including Chen Yun, Li Fuqun, and Li Xiannian, all advocated improved price incentives.[21] They proposed not across-the-board price increases, but specifically targeted increases for crops whose prices were low relative either to previous levels or to production costs. Tung oil, rapeseed, peanuts, tea, silk, and pigs were mentioned most prominently. The case of pigs was particularly interesting because the rising cost of production was attributed by Chen Yun to government policy. The by-product of rice milling, bran, was traditionally the most important source of pig fodder, but since the socialization of grain processing peasants were no longer allowed to retain the bran but had to buy it back from the state, raising the cost of pig production.[22]

The Eighth Party Congress substantially modified the policies the party had pursued in the countryside over the previous year. The party Central Committee and the State Council, the highest organ of governmental authority, in mid-September jointly issued a directive calling for a formal end to production planning and the return to a system in which cropping decisions would be constrained only by the requirement of fulfilling delivery quotas.[23] On September 29, two days after the congress adjourned, the State Council promulgated a directive raising the price of rapeseed relative to wheat by 50 to 70 percent, depending on the particular geographic location. That was followed, a few months later, by a 14 percent increase in the price the state paid for hogs purchased from the peasantry.[24] Price adjustments, the reduction in the number of sown area and output targets transmitted to producers, and the reopening of rural markets expanded peasant income-earning opportunities and, by reducing the number of planning constraints they faced, may have contributed to improved productivity.

The course of events after the reopening of rural markets in the summer of 1956 revealed, however, a persistent, fundamental cleavage between state and peasant interests, one which was to reemerge in the 1960s and in the late 1970s. In short, rural markets undermined the ability of the state to purchase agricultural products at fixed prices in order to feed the urban population, supply agricultural raw materials to state-controlled light industry, and meet export demand. When markets were reopened in the summer of 1956 the state intended that they be limited to what were referred to as native and subsidiary products, items that, in short, were not subject to quota targets for delivery to the state.[25] It proved difficult, however, to exclude cereal, oil, and fiber crops from these markets, which at that time numbered about 40,000.[26] Peasants preferred to sell these items on free markets, where prices were substantially higher than those paid by the state. Thus in 1956, although grain production rose 5 percent, state grain taxes and purchases fell from 43.0 to 41.7 million tons,[27] and rural private-market sales increased from 2 to 3 million metric tons in 1954–55 to 4 to 6 million metric tons in 1956–57.[28]

Concern that the free market was undermining state procurement was reflected in an article published in the journal of the State Planning Commission in December 1956.[29] While praising the rural free markets in principle, the operative recommendations were to forbid any transactions in grain, oil seeds, and cotton on these markets and to institute formal price controls on other, nonspecified commodities. Eight months later, in August 1957, regulations closely following these recommendations were promulgated by the State Council.[30] Sale of cereals, oil-bearing crops, and cotton in rural markets formally was prohibited. Surplus grain could be sold only directly to state agencies or in state-controlled grain markets, which had replaced traditional grain markets in cities and market towns in 1953. Even though free-market sales were eliminated and a major campaign launched to increase deliveries to the state at fixed prices, government procurement (including tax grain) fell to 39.8 million tons, although grain production posted a marginal 1 percent increase compared to 1956.[31]

The Great Leap Forward reveals perhaps more than any other single period the detrimental effect of party policies on the peasantry. The history of the period is complex and only a few aspects can be treated here. Again the undermining of production incentives through the formation of units that at their largest averaged over 5,000 families and the pursuit of highly egalitarian distribution of collective output is relatively well known and falls outside the scope of the analysis presented here.

The best-known state policies detrimental to peasant livelihood during the late 1950s were the curtailment of rural markets and the widespread elimination of private plots. In an acceleration of the trend begun in the fall of 1957, when rural markets came under increased state pressure, rural markets in most regions of China were closed in the fall of 1958 as the movement to form communes got underway. As in the winter of 1955–56, market closure reduced the opportunities of peasants to earn income from household sideline activities. The negative effects on income presumably were larger than in 1955–56 since private plots, an important source of products not only for self-consumption but also for sale on rural markets, were eliminated.

The closure of rural markets was undoubtedly a preconditon for a less well-known aspect of party policy that was detrimental to peasant interests—a vast acceleration in the forced deliveries of cereals at fixed prices. In 1958 state purchases increased substantially over 1957, presumably at least in part since local party cadres reported enormous increases in cereal production in the wake of the formation of communes.[32] The officially reported cereal output for 1958 was 375 million metric tons, almost double the 195 million tons of 1957.[33] The increases in production turned out to be largely fictitious, but the increases in procurement were very real. As early as the spring of 1959 there were already reports of significant local food shortages, and as the year progressed conditions only worsened. Cereal production fell significantly while state purchases of cereal crops remained far above the pre-Leap levels.[34] The situation was only worse in 1960. Production reached a trough of 144 milion tons, 51 milion tons below the level of 1957.[35] Forced deliveries dropped compared to 1958 and 1959 but remained in excess of 1957.[36]

The consequence of these trends was devastating for the Chinese peasantry. The national mortality rate began to rise significantly as early as 1959. In 1960 the mortality rate reached a peak of 25.43 per thousand, more than double the average rate of 11.39 prevailing in 1956–58. In 1961 the rate declined but was still higher than normal.[37] Cumulatively in 1959–1961, the number of excess deaths, i.e., over and above the number that would have occurred with prefamine levels of mortality, exceeded ten milion. Although the incidence of increased mortality has not been revealed in Chinese sources available to me, famine and famine-related deaths probably were disproportionately rural. The best evidence for this hypothesis comes from an internal party history that shows while on average cereal consumption declined 20 percent from 203.0 kilograms in 1957 to 163.5 kilograms in 1960, almost all of the reduction in consumption was borne by the peasantry.[38] Their average consumption in 1960 declined to 156.0 kilograms, almost 25 percent below the level of 1957, while urban consumption of cereals on average fell less than 2 percent. In short, the Chinese Communist Party chose to insulate urban consumers from the effects of a famine of unprecedented magnitude in this century by stepping up forced deliveries from the peasantry and drawing down state-controlled stockpiles to feed urban residents.[39]

Some have suggested that the major responsibility for the extraordinary loss of life during these years lies with local party cadres who sought to conceal their initially inflated cereal-output reports by subsequently failing to report serious food shortages. I find this argument wholly unconvincing. First, two of the highest ranking members of the Politburo made extensive visits to rural areas in the winter of 1958–59 and reported in detail on rural conditions they observed. Peng Dehuai, visiting in Hunan, found an enormous gulf between the conditions he observed and those he expected on the basis of the increases in agricultural production that had been reported to Beijing. He found that "the masses were in danger of starving"; not waiting until his return to Beijing to report his findings, he sent an urgent cable to the Central Committee warning that the 1958 production figures that had been submitted previously were overstated.[40] Chen Yun carried out his investigations in the spring of 1959 in Henan, whose provincial party secretary, Wu Zhipu, had been credited by Mao as having established the first commune in 1958. Chen, too, challenged the accuracy of the figures being reported to Beijing but reserved his most penetrating questioning of provincial officials to the topic of the balance between urban and rural grain supplies. When asked what they would do if they found urban supplies falling short, provincial party officials, apparently deluded by their own inflated output reports, blithely responded that they would simply step up extractions from the peasantry.[41]

The great tragedy of the Chinese peasantry at this time was that officials, such as Chen Yun and Peng Dehuai, who challenged the success of Mao's Great Leap strategy were stripped of influence or formally purged. The increasingly charged political atmosphere leading up to the Lushan Plenum made rational policy making impossible. Not until many million more peasants died was the folly of the Great Leap Forward explicitly admitted by the party and more rational policies adopted.

In 1961–62 state policies became substantially more favorable toward rural areas. The state simultaneously increased the fixed price paid for cereals by 25 percent, the first significant price increase since the introduction of compulsory deliveries in the fall of 1953, and reduced the delivery quotas to below the levels prevailing during the first plan.[42] Other crop prices were adjusted as well, in both 1961 and 1962. Responsibility for production decisions was devolved from communes to production teams and in many regions to smaller groups or households, and private plots generally were restored. Equally important, rural markets were reopened, in some places as early as the summer and fall of 1959, restoring income-earning opportunities and increasing the possibility of specialized production. Initially few in number and restricted by government regulation, the revival of these markets appears to have been a key element in the restoration of the rural economy in the wake of the Great Leap Forward debacle. By 1961 40,000 rural markets had been reestablished, marking a restoration of the level of the mid-1950s.[43] Initially, as in the summer of 1956, transactions in grain, edible vegetable oil seeds, and cotton were prohibited, but in 1962 provinces were authorized to allow these products to enter rural markets once delivery quotas to the government had been fulfilled.

Recovery of agricultural output was facilitated, as well, by returning to a more rational interregional pattern of cropping based on comparative advantage. Chen Yun, who by 1960 had returned to play a leading role in economic policy formulation, forcibly argued the necessity of abandoning the concept of self-sufficiency on the provincial or commune level, which had been an integral part of the Great Leap Forward.[44] In place of local self-sufficiency the party endorsed a policy of accentuated development of cereal production and marketing in "high- and stable-yield grain areas" and specialized production of cotton, sugar cane, and other noncereal crops. The objectives of this strategy were to increase the efficiency of resource allocation in agriculture and increase the rate of marketing.

Although the policy mix adopted in the first half of the 1960s was successful in raising agricultural output and productivity, it gradually was abandoned after 1966. Prices of most agricultural commodities were frozen in a pattern that was not to change until 1978. Producers were less able to pursue comparative-advantage cropping because the state increasingly constricted the scope of rural markets and reduced the interregional flows of agricultural commmodities that facilitated specialized production. Self-sufficiency for each locality gradually was reinstituted as a major policy objective.

Although Mao's vision of self-sufficient rural communes, each with its own food-processing plants, small-scale industries, and social-service system, had been abandoned in the early 1960s, Mao remained profoundly antagonistic to the concept of specialized production based on comparative advantage. In his notes (written in 1960) on the Soviet textbook *Political Economy* Mao rejected the concept of international comparative advantage.[45] His critique, however, extended to China's domestic economy as well. Mao was insistent not only that China as a whole should be self-sufficient in cereals, but that it was dangerous for any area of China to be dependent on grain supplied by other provinces.

Although the policies that led to agricultural recovery in the first half of the 1960s were based on exploiting existing comparative advantage, Mao continued to press for self-reliance, for example, at a Central Committee Work Conference in 1964.[46] In a March 1966 speech to an enlarged meeting of the Politburo of the Central Committee, Mao took up the topic with greater urgency.[47] The dependence of North China on grain from the south was the specific target of his attack. There were several reasons Mao's ideas were apparently less easily ignored at this time. First, Mao was then moving to reestablish his predominance over China's political system through a movement that would come to be known as the Great Proletarian Cultural Revolution. His withdrawal from policy making, which had followed the failure of the Great Leap, came to a rather abrupt end. Second, the concept of local self-sufficiency was considered as one element of a military strategy postulated on a war of attrition fought by decentralized, self-sufficient regional forces.[48] Whether the leadership was more concerned with a spillover of the war in Southeast Asia into China or was anticipating a Soviet invasion in the north is not clear. But the creation of regions that could survive even if cut off from neighboring provinces and regions became an official policy goal. That goal was probably reinforced after 1968 when military clashes occurred on the Sino-Soviet border.

While the slogan "take grain as the key link" was heard increasingly after the mid-1960s, it is difficult to judge the degree to which self-sufficiency was pursued as a policy objective or to judge the extent to which this policy reduced income-earning opportunities. However, there is substantial evidence that in the Cultural Revolution period (from about 1966 to 1978), trade and specialized production based on comparative advantage were curtailed systematically. That reduced peasant income-earning opportunities, particularly for regions that were relatively specialized in the production of nongrain crops and depended on trade to meet their major consumption needs.

The decline in trade, specialization, and income-earning opportunities is reflected in both national data and in the histories of agricultural development of many individual regions.

One aggregate reflection of decreased commercialization was a decline in the rate of marketing of cereal crops. Not only was private trade suppressed, but while cereal production rose from 214.0 milion tons in 1966 to an average of 306.5 milion tons in 1976–78, state purchases of cereals rose by an average of only 2.9 million metric tons.[49] By 1976–78 the marketing rate for grains was lower than during the mid-1950s, although per capita production had more than recovered to the peak production levels of the 1950s.

A second reflection of the self-sufficiency policy was a sharp decline in interprovincial transfers of cereals. By 1978 total interprovincial transfers of cereals were only 2.05 milion tons, less than a third of the average level in 1953–56, 7 to 7.5 million tons.[50] Interregional cereal trade as a share of production actually declined from 4.7 percent in 1953–56 to .8 percent in 1978. In absolute terms transfers in 1978 were less than half the 4.7 million tons shipped interprovincially in 1965.[51] If rice destined for international markets is excluded from these transfer figures, the decline in interregional trade is even more dramatic, .3 percent in 1978 compared to 3.6 percent in 1953–56 and 1.8 percent in 1965.[52]

Although it could be argued that the reduction in transfers reflects success in achieving the stated goal of local self-sufficiency, the strategy was a costly one from the point of view of the Chinese peasantry. First, it resulted in a skewed pattern of agricultural development. Cereal output growth (excluding soybeans) was fairly rapid, in excess of 3 percent per year, and by the mid-1970s per capita production had recovered to the peak level achieved prior to the Great Leap Forward. Other food crops, however, did not recover to the levels of per capita output of the 1950s. Per capita consumption of vegetable oils declined by a third. Per capita production of soybeans was half the level of 1957. On average, per capita caloric consumption in 1977–78 was probably somewhat below the level of 1957, and rural consumption appears to have declined vis-à-vis urban.[53]

Second, the costs of increased inefficiency arising from less specialization and the imposition of ill-conceived cropping patterns, such as triple cropping, are reflected in the sharp increase in nonlabor costs as a percentage of the gross value of output. These costs, which include outlays for feed, fertilizers, and other purchased inputs, were stable at 26–27 percent of output value in the 1950s, when individual producers were able to make their own decisions about cropping patterns and levels of input use. Costs presumably rose in the early 1960s because of the extreme food shortages, but by 1965 costs had fallen to 27.3 percent of output. Again that reflects the more modest state interventions in crop production decisions and state policies that facilitated specialized production. But sometime after 1965 costs began to rise, to 31.9 percent by 1974, and 33.6, 35.5, and 35.7 percent in the subsequent three years.[54]

The implications of that rise in costs are important in shaping our understanding of the 1965–1977 period. The official data on gross value of output show that agriculture (in constant prices) grew at the respectable rate of 3.7 percent.[55] But that measure, one frequently cited by Chinese sources, does not reflect the increasing costs of producing over that period. On a value-added basis, that is, net of the costs of purchased inputs, farm output grew only 3.2 percent.[56] In per capita terms, ignoring increased costs, it appears that output growth was over 1.5 percent per annum. Taking the growth of costs into account shows that per capita agricultural output grew only 1 percent per annum. Thus the gross value measure overstates the rate of per capita growth by half.

On balance, then, agricultural growth between 1965 and 1977–78 was skewed and, on average, increasingly costly. These data, based on national averages, are somewhat abstract and are perhaps placed in perspective by examining how state marketing policies shaped the income-earning opportunities of one specific group of Chinese peasants, sugar cane producers in Fujian province.[57]

Comparative advantage, in this case determined largely by climate, is the basis for sugar cane cultivation in Fujian. Specialized production, of course, is dependent on distant markets for sugar as well as nonlocal sources for cereals. Sugar produced in Fujian was sold throughout the Chinese empire, as well as in Southeast Asian markets, as early as the Ming dynasty. Central coastal Fujian, where sugar cane, mulberry leaves (for silkworm production), and other nongrain crops were grown, depended on cereal imports from the Yangze delta, the Canton delta, and the central Gan River valley as early as the twelfth century. Dependence on external sources of rice persisted throughout the eighteenth and nineteenth centuries. Specialization in sugar production may have increased in the latter half of the nineteenth century when steamships entered the coastal trade, reducing costs of transport to ports such as Shanghai and Tianjin on the coast in Central and North China, respectively, as well as to major interior cities such as Hankou on the Yangze River. Sugar was even shipped to Newchang (Yingkou) in Liaoning after it was opened in 1861 and traded for beancake, which was shipped to South China where it was used as a fertilizer, especially in production of sugar cane.

Specialized production of sugar cane and the concomitant trade that facilitated that production pattern were maintained and even accelerated during most of the 1950s. Favorable price incentives and the restoration of trading after the war and civil war contributed to rapid recovery of the sugar cane area, which had fallen by 1949 to just over a third of the area cultivated during the 1930s. By 1957 sugar cane area and yields in Fujian were both 50 percent above the peak prewar year,

1936. The absolute yield was 49.9 tons of raw cane per hectare, the highest of any province in China.

In part Fujian's performance was achieved by increased specialization in sugar cane production within the province. Cane production in the 1950s was increasingly concentrated in five counties in Jinjiang prefecture. By 1957 four of these counties, Xianyou, Putian, Nanan, and Tongan cultivated half of the sugar cane area and produced two-thirds of the province's output. Xianyou was the leading producer, as it had been in the thirteenth century. Yields there in 1957 were 84 tons per hectare, allowing the county to produce 45 percent of the raw cane and 50 percent of the refined sugar of the entire province. Cane was cultivated on 17 percent of the county's land in 1957, about 10 times the share of land sown to cane in the province and 100 times the share of national sown area allocated to cane.

Although I have been unable to locate data on exports of sugar from Jinjiang prefecture or Fujian as a whole for any year during the 1950s, only a small share of sugar production could have been consumed within sugar producing regions of the province. Most of the cane was refined in state-owned refineries and transported to be consumed elsewhere. In turn, coastal areas of Fujian, where 92 percent of the province's cane was grown, depended on external sources of grain throughout the first five-year plan.

In the crisis years of the late 1950s and early 1960s sugar production in Fujian dropped sharply, presumably because the national collapse of grain output reduced supplies shipped to sugar producers in Fujian. In Xianyou, for example, in 1962, when national grain production had begun to recover, sugar-sown area was down by almost half compared to 1957 and yields were off 60 percent, presumably because available supplies of fertilizer and other inputs had been shifted into cereal production. However, output recovered quickly in response to the improved marketing arrangements and specialization policies adopted in 1961 and 1962. By the 1966–67 refining season, output of sugar in Fujian was almost twice the level of 1957, and exports to the rest of China reached 55 percent of production.

In the ensuing decade,when local self-sufficiency in cereals was pursued, sugar cane output declined as opportunities to market refined sugar and to purchase cereals diminished. In a manner similar to the crisis of the late 1950s and early 1960s, current inputs were reallocated to grain fields and cane yields in a decade fell to 30 percent below the level of 1957. By 1976, output of refined sugar and shipments out of the province were 40 and 80 percent below the levels of 1967.

Not surprisingly, peasant income in sugar-producing regions fell substantially as a result of the state's curtailment of their income-earning opportunities. By 1975, per capita consumption of cereal by sugar-producing peasants in Fujian averaged only 150 kilograms. In Xianyou the level was 145 kilograms. That was very far below the official national poverty standard of 200 kilograms of rice (unprocessed) that prevails in South China.

Fujian's declining production was arrested and turned around within a short time after 1976 by the reintroduction of marketing opportunities. The restoration of comparative-advantage cropping and increased interregional trade in Fujian is unusual because it predates by three or four years the ratification of a similar policy for producers of other noncereal crops, such as cotton. It also preceded by several years the restoration of trading opportunities in sugar-producing regions in Guangdong and Guangxi. Curiously it has not received very much press attention within China, perhaps since it is an example of a successful policy adopted prior to the Third Plenum of the Central Committee, a meeting which is credited with being the font of agricultural revival in China.

Under the system announced in 1976, the central government began to supply Fujian annually with 100,000 tons of grain, which the province was to sell to specialized producers of cane. For every thousand tons of cane sold to the state in excess of a base target of 984,000 tons, production teams were guaranteed the right to purchase 125 kilograms of unhusked grain at the then prevailing government quota price for cereals. The access of producers to both fertilizers and refined sugar also was strengthened. For every ton of cane sold to the state, the quantity of chemical fertilizer that peasants were guaranteed the right to purchase was raised from 15 to 25 kilograms. They were also guaranteed the right to purchase 10 kilograms of refined sugar for each ton of cane sold. Sugar-producing peasants could trade back any portion of this sugar for grain at a ratio of one to two.

The easing of the artificial constraints imposed by the government over a decade had a remarkable effect on sugar cane production and peasant income in Fujian. Between 1976 and 1981 production of cane sugar more than tripled, while yields more than doubled. More interestingly, exports of refined sugar from the province went up more than sevenfold to reach 180,000 tons by 1980. During the first five years this system was in place, peasants engaged in sugar cane production purchased outright or bartered with excess sugar above their own personal consumption needs (from the 10 kilograms of sugar they were entitled to buy) 717,500 tons of grain. Per capita cereal consumption of sugar-producing peasants reached 250 kilograms and 270 kilograms in the whole province and Xianyou county, respectively, in 1979.

The most important point to be understood about the resurgence of cane production and rising peasant income and consumption in Fujian's cane-producing regions is that they were achieved without any significant increase in the price of sugar cane relative to rice or other cereals. Between 1975 and 1979 the purchase price of rice and sugar cane rose 20.3 and 22.5 percent, respectively. The increased cane-sown area and the rise in yields are the result of the reduction of previously imposed artificial constraints on the markets for agricultural goods (cereals and sugar cane) and on industrial inputs to farming, not an increase in the relative price of sugar cane or special production subsidies. Thus, the large increase in income and consumption may be thought of as a pure allocative efficiency gain achieved through a restoring of marketing and the pursuit of comparative-advantage cropping.

Sugar cane producers in Fujian are not unique. Since 1978, state rural policy in general has become less inimical to peasant livelihood. The state has raised the prices it pays for all agricultural crops and has sanctioned the reopening of rural markets. Moreover, the improvement in peasant income opportunities that can be achieved through comparative-advantage cropping, made possible through state redistribution and marketing of cereals over greater geographic areas, has been extended to regions such as Hainan Island and Shaanxi province and to specialized cotton producers in several provinces through specially guaranteed external supplies of cereals.

While the contribution of these policies to raising farm productivity and income is increasingly known, the state confronts a number of dilemmas that have led it to continue and in some ways to increase the constraints under which peasants must labor. The modification of the more liberal policies introduced in 1978–79 is evident in state policy toward rural markets, on prices, and on interregional specialization.

As in 1956–57 and again in the first half of the 1960s, the expansion of rural marketing opportunities has increased the cost of state cereal procurement. In the reforms of 1979 the basic quota price for cereals was raised 20 percent and the premium price for overquota deliveries increased from the old level of 30 percent to 50 percent. Despite that substantial increase, peasants actually sold only a moderately higher portion of their cereal output to the state in 1979, 1980, and 1981.[58] Higher prices effectively reduced the tax burden implicit in quota and overquota sales, but the small increase in sales suggests that some tax burden remains. The only increase in voluntary deliveries was in the category of negotiated purchases for which the state pays a substantially higher price, linked closely to the rural market price. For rice in 1980 the price necessary to induce voluntary deliveries was more than twice the quota price and more than 50 percent above the overquota price. For wheat the differentials were about 80 and 20 percent, respectively.[59]

Thus the revival of rural markets has provided peasants with an attractive alternative to selling to the state at fixed prices. To reduce this advantage the state has placed an increasing number of restrictions on rural market transactions. Sales of raw cotton and handicraft yarn and cloth are, as in the past, legally precluded.[60] Sales of grain, edible vegetable oils, and other commodities purchased by the state are not allowed until both quota and overquota delivery targets have been fulfilled. But by 1980 there were 37,890 rural markets,[61] so peasants sought to evade this restriction by selling output in nearby markets where trade in grain had been sanctioned by virtue of the prior fulfillment of delivery quotas. Thus the state placed further restrictions on these markets. First, collective units, which presumably had better access to transportation to move their grain to nearby markets, were prohibited from selling any grain in rural markets.[62] And sales by individuals were legally restricted to producers. Resales by middlemen, identified as "profiteers" or "speculators," were prohibited. Even individuals were prohibited from selling any products subject to unified sale beyond markets adjacent to their own. Moreover, the state has placed price controls on commodities sold in most markets.[63]

These actions effectively reduce income-earning opportunities of the peasantry and force their transactions to be more concentrated with state agencies than they would voluntarily choose. Simultaneous with its efforts to make rural markets less attractive, the state has placed a ceiling on prices it will pay peasants for products sold to the state. No less an authority than Party Chairman Hu Yaobang has proclaimed that improvements in peasant income must be the consequence of productivity growth, not higher prices.[64] That policy, in turn, is engendered by the apparently unshakeable commitment by the party to supply cheap staple commodities, notably cereals and edible vegetable oils, to urban residents and a smaller number of state employees who live in rural areas and are included within the rationing system. Because the retail prices of cereals sold through the rationing system have not risen significantly since 1952, while purchase prices have more than doubled, the state incurs huge financial losses in cereal and vegetable oil transactions. In 1981 these losses were in excess of 25 billion yuan, equivalent to one-fourth of budgetary revenues at all levels of government or one-third of the wage bill for all state employees.[65] In the short run a choice has been made to cap the growth or reduce the magnitude of these subsidies not by increasing the price of food to urban consumers, which in real terms costs substantially less than in the 1950s, but by freezing or perhaps even reducing the prices paid to farmers.

While peasants have been told to depend increasingly on productivity growth to achieve higher levels of income, the state has moved to reduce peasant opportunities to become more productive by freezing, for a period of three years beginning in 1982, interprovincial flows of agricultural commodities.[66] Since private transactions are limited in both volume and geographic scope, the state is the sole participant in interregional commodity flows. Freezing the flows of commodities inhibits the development of comparative-advantage cropping since, as in the case of sugar cane production discussed above, it is heavily dependent on state-controlled interregional commodity flows.

While the rural market, price, and interregional trade policies in 1978 and 1979 had initially favorable effects on agricultural output and productivity, as well as on farm income, the party has sought to modify these policies in more recent years. Whether these modifications are temporary or reflect the ascendancy of leaders whose view of development focuses primarily on capital-intensive industrialization at the expense of agriculture will be evident from the future course of policy.

## Notes

1. Conrad Brandt, *Stalin's Failure in China, 1924–1927* (Cambridge: Harvard University Press, 1958); Lucien Bianco, *Origins of the Chinese Revolution, 1915–1949* (Stanford: Stanford University Press, 1971).

2. Joel S. Migdal, *Peasants, Politics, and Revolution: Pressures Toward Political and Social Change in the Third World* (Princeton: Princeton University Press, 1974). For one of the strongest variants of this view see Ralph Thaxton, *China Turned Rightside Up: Revolutionary Legitimacy in the Peasant World* (New Haven: Yale University Press, 1983).

3. Charles R. Roll, *The Distribution of Income in China: A Comparison of the 1930s and the 1950s* (New York: Garland, 1980). Not all of those who attribute the Chinese Communist Party's rise to power to their shift to a rural-based strategy after 1927 accept the view that the party continued to serve peasant interests after 1949. Bianco, for example, states that after 1949 the party "used the peasants to force the progress of history, without wasting much time on the narrow-minded aspirations of the rural masses" (p. 77).

4. Farm output value (measured on a net basis and in constant prices) grew at an annual rate of 5.9 percent between 1952 (by which time recovery from wartime damage was largely complete) and 1957. Shigeru Ishikawa, *National Income and*

*Capital Formation in Mainland China: An Examination of Official Statistics* (Tokyo: Institute of Asian Economic Affairs, 1965), p. 56. The most careful study of the increase in peasant consumption during these years is David Denny, "Rural Policies and the Distribution of Agricultural Products in China," Ph.D. diss., University of Michigan, 1971.

5. Thomas P. Bernstein, "Leadership and Mass Mobilisation in the Soviet and Chinese Collectivisation Campaigns of 1929–30 and 1955–56: A Comparison," *The China Quarterly* 31 (July-September 1967):1–47; Vivienne Shue, *Peasant China in Transition: The Dynamics of Development Toward Socialism* (Berkeley: University of California Press, 1980).

6. Dwight H. Perkins, "The Central Features of China's Economic Development," in *China's Development Experience in Comparative Perspective*, ed. Robert F. Dernberger (Cambridge: Harvard University Press, 1980), pp. 120–50.

7. Dwight H. Perkins, "Constraints Influencing China's Economic Performance," in *China: A Reassessment of the Economy*, papers submitted to the Joint Economic Committee, Congress of the United States (Washington, D.C.: GPO, 1975), p. 350.

8. Harvey Leibenstein, "Allocative Efficiency vs. X-Efficiency," *American Economic Review* 58, 2 (June 1966):392–415.

9. Government Administrative Council, "Directive Implementing Planned Purchase and Planned Supply of Grain," *New China Monthly* 4 (1954):158–59.

10. The limited character of Soviet financial assistance and the critical importance of Chinese agricultural exports to the Soviet Union are discussed in Alexander Eckstein, *Communist China's Economic Growth and Foreign Trade* (New York: McGraw Hill, 1966).

11. Yujiro Hayami, K. Subbarao, and Keijiro Otsuka, "Efficiency and Equity in the Producer Levy of India," *American Journal of Agricultural Economics* 64, 4 (November 1982):655–63.

12. Government Administrative Council, "Directive Implementing Planned Purchase."

13. Government Administrative Council, "Provisional Measures for the Management of Grain Markets," *New China Monthly* 4 (1954):159–60.

14. Alexander Eckstein, *China's Economic Revolution* (Cambridge: Cambridge University Press, 1977), p. 117; Shigeru Ishikawa, "Resource Flow Between Agriculture and Industry: The Chinese Experience," *The Developing Economies* 5, 1 (March 1967):42.

15. The limited quantities of cotton retained by peasants in cotton-producing regions were for the purpose of padding for bed quilts and winter jackets.

16. Even in the late 1970s, when the price the state paid peasants for raw cotton was 50 percent higher than in the 1950s, the cotton-textile industry was the source of more than 10 percent of the revenues of the unified state budget. Ministry of Textile Research Office, "China's Textile Industry," in *Annual Economic Report of China 1981*, ed. Xue Muqiao (Hong Kong: Modern China Culture Company, 1982), vol. 4, p. 48.

17. State Council, "Provisional Measures for the Supply of Fixed Quantities of Grain to Cities and Towns," *New China Monthly* 9 (1955):163–64.

18. Fei Xiaotong, "A Revisit to Kaihsienkung," *New View* 11 and 12 (1957), in *Fei Hsiao-t'ung, The Dilemma of a Chinese Intellectual*, trans. James P. McGough (White Plains, NY: M.E. Sharpe, 1979), pp. 39–74.

19. Kenneth R. Walker, "Organization for Agricultural Production," in *Economic Trends in Communist China*, ed. Alexander Eckstein, Walter Galenson, and T. C. Liu (Chicago: Aldine, 1968), p. 424.

20. Chen Yun, *Selected Manuscripts of Comrade Chen Yun, 1956–1962* (Beijing: People's Publishing House, 1981), p. 29.

21. Li Fuqun, "Speech by Comrade Li Fuqun," in *Eighth National Congress of*

*the Communist Party of China* (Beijing: Foreign Languages Press, 1956), vol. 2, pp. 288–303; Li Xiannian, "Speech by Comrade Li Xiannian," in *Eighth National Congress of the Communist Party of China* (Beijing: Foreign Languages Press, 1956), vol. 2, pp. 206–24.

22. Chen Yun, *Selected Manuscripts*, p. 15.

23. Chinese Communist Party Central Committee and State Council, "Directive Concerning Strengthening Production Leadership and Organization of Construction in Agricultural Producer Cooperatives," *People's Daily,* September 13, 1956, pp. 1, 3.

24. State Council, "Directive Concerning the Increased Procurement Price for Rapeseed," in *A Compendium of Materials on Prices in Shanghai Before and After Liberation*, comp. Shanghai Economic Research Institute, Chinese Academy of Sciences, and Economic Research Institute, Shanghai Academy of Social Sciences (Shanghai: Shanghai People's Publishing House, 1958), pp. 594–95; "Directive Concerning the Readjustment of Pig Procurement and Sales Prices," in *Compendium of Laws and Regulations of the People's Republic of China*, comp. State Council Bureau of Legal Affairs (Beijing: Legal Publishing House, 1957), vol. 5, pp. 181–82.

25. The fundamentally different character of these markets is also reflected in their name "free markets" (*ziyou shizhang*) as opposed to the state-controlled markets for grain (*guojia liangshi shizhang*), which had replaced traditional grain markets in cities and market towns (*chengshi he jizhen zhongti liangshi jiaozhang isuo*). By contrast, when markets were revived in the first half of the 1960s and again in the late 1970s, they were referred to as "rural markets" (*nongcun jishi*), not "free markets."

26. This number is a rough estimate based on G. W. Skinner's observation that there were 42,900 standard rural markets in China in the late 1940s and that there were 40,000 rural markets again in the early 1960s and 38,000 rural markets in 1980. G. W. Skinner, "Marketing and Social Structure in Rural China," *Journal of Asian Studies* 24, 2 (February 1965):228. Sources for the number of markets in the early 1960s and in 1980 are cited below in notes 43 and 61.

27. Grain output data are contained in Agricultural Yearbook Compilation Commission, *Chinese 1980 Agricultural Yearbook* (Beijing: Agricultural Publishing House, 1981), p. 34. Grain tax and procurement data are from Statistical Work Data Office, "The Basic Situation of Unified Purchase and Sale of Foodgrains in China," *Statistical Work* 19 (1957):31–32, 28. These procurement data, attributed to the State Statistical Bureau, differ somewhat from a more recent source that shows that procurement fell from 47.54 to 40.22 million metric tons between 1955 and 1956. State Statistical Bureau, *Chinese Statistical Yearbook 1981* (Beijing: Statistical Publishing House, 1982), p. 341. Both series are said to be measured in terms of "trade weight," but they may vary for a number of reasons. Procurement data are variously reported on either a calendar year or a grain year basis (frequently without specifying which is being used), and in addition to amounts purchased by the Ministry of Food and its local agencies may include market purchases by state organizations, military units, mass organizations, schools, and enterprises.

28. Alexander Eckstein, *China's Economic Revolution*, p. 117; Shigeru Ishikawa, "Resource Flow Between Agriculture and Industry," p. 42.

29. Sun Imin, "Strengthen Leadership Over the Free Market," *Economic Planning* 12 (1956) in *Extracts from China Mainland Magazines* 77 (April 19, 1957) (Hong Kong: U.S. Consulate General):14–20.

30. State Council, "Regulations on the Restrictions on Certain Agricultural Products and Other Commodities Which Are Subject to Planned Purchase or Unified Purchase by the State," *New China Semi-Monthly* 18 (1957):207–208.

31. Again there are conflicting reports on procurement in 1957. The number cited in the text is that given by Ministry of Agriculture Policy Research Office, *China's Basic Agricultural Situation* (Beijing: Agricultural Publishing House, 1982), p. 30, and by Liang Xiufeng, "Correctly Draw the Lessons of Experience of the Three Years

of the Great Leap Forward and the Five Years of Readjustment,'' *A Collection on Agricultural Economics* 4 (1981):20. *Chinese Statistical Yearbook 1981*, p. 341, shows procurement rose from 40.22 million metric tons in 1956 to 45.97 million metric tons in 1957. These numbers, however, appear to be part of a series that includes sale of grain in rural markets, direct sales by farmers to city dwellers in urban markets, and perhaps sale by some government units other than grain purchase agencies.

32. Because of the conflicting data on procurement the magnitude of the increase is unclear. The increase may have been as much as 40 percent, based on a figure of 39.8 for 1957 (sources in note 31) to 55.7 million tons in 1958. The latter is from Alexander Eckstein, *China's Economic Revolution*, p. 17. The increase in procurement may have been as little as 13 percent, based on the procurement data reported in the 1981 statistical yearbook.

33. The figure of 375 million metric tons was first reported in the December 1958 Communique of the Sixth Plenum of the Eighth Central Committee, see *People's Handbook* (Beijing: Dagong Publishers, 1959), pp. 37–38, for text. Following the Lushan Plenum in the fall of 1959 the output figure was revised downward to 250 million tons, and in the 1980 Agricultural Yearbook the 1958 harvest was given as 200 million tons.

34. Cereal output in 1959 was 170 metric tons, down 25 million tons or 13 percent from 1957. Chinese Agricultural Yearbook Compilation Commission, *China's 1980 Agricultural Yearbook* (Beijing: Agricultural Publishing House, 1981), p. 34. Liang Xiufeng states that 1959 procurement was 40 percent over that of 1957, rising from 48.2 to 67.5 million tons (in original weight; from 39.8 to 55.9 in terms of trade weight). *Chinese Statistical Yearbook 1981* also shows a rise of 40 percent, but from 45.97 to 64.12 millions tons (both in trade weight). Even taking into account the grain resold in the countryside, the net extraction rate in 1959 reached an historic peak level of 28 percent, more than 50 percent greater than the average net rate of the First Plan. Yang Jianpai and Li Xuezeng, "On the Historical Experience of the Relations Between Agriculture, Light Industry and Heavy Industry in China," *Social Sciences in China* (Chinese edition) 3 (1980):36.

35. *China's 1980 Agricultural Yearbook*, p. 34.

36. Liu Suinian, "The Proposal and Implementation of the Eight-Character Policy of Readjustment, Consolidation, Filling-Out and Raising Standards," *Research on Party History* 6 (1980):23, says 1960 procurement was 51.05 million tons, but this is probably measured in original weight. In terms of trade weight that would be about 42.5 million metric tons, 2.7 million tons more than my estimate of 39.8 for 1957, but down substantially from my estimate of procurement in 1958 (55.7 million metric tons) and 1959 (55.9 million metric tons). The *Statistical Yearbook 1981*, p. 341, shows a lower figure of 46.54 million tons for 1960, less than a million tons more than its figure for 1957. But the 1960 level is very far below the level of 1959, which was 64.12 million tons, according to the yearbook.

37. Zhang Huaizu et al., *A General Survey of Population Theory* (Zhengzhou: Henan People's Publishing House, 1981), p. 83.

38. Liu Suinian, "The Eight-Character Policy," p. 24.

39. Chen Yun, *Selected Manuscripts of Chen Yun*, pp. 122–23. The drawdown on state inventories is analyzed in Nicholas R. Lardy and Kenneth Lieberthal, *Chen Yun's Strategy for China's Development: A Non-Maoist Alternative* (Armonk, NY: M.E. Sharpe, 1983).

40. Li Rui, "Reading Peng Dehuai's 'Account,' " *Reading* 4 (1982), trans. in Foreign Broadcast Information Service, *China Report,* April 16, 1982, pp. K5-K6.

41. Deng Liqun, *Study How to Do Economic Work from Comrade Chen Yun* (Beijing: Chinese Communist Party Central Party School Publishing House, 1981), pp. 54–55.

42. Liu Suinian, "Eight-Character Policy," pp. 23–24, 28.

43. Ho Cheng and Wei Wen, "On Peasant Periodic Market Trade," *Economic Research* 4 (1962):12.

44. Chen Yun, *Selected Manuscripts of Chen Yun,* pp. 112–13.

45. Commenting on the text statement that "Each country should develop its own manpower and material resources to develop its own most advantageous natural and economic conditions," Mao wrote flatly "that is not a good idea." "Reading Notes on the Soviet Text Political Economy," in *Long Live the Thought of Mao Tse-tung* (Taipei, 1969), in *A Critique of Soviet Economics,* trans. Moss Roberts (New York: Monthly Review Press, 1977), p. 102.

46. Kenneth Lieberthal, *A Research Guide to Central Party and Government Meetings in China 1949–1975* (White Plains, NY: International Arts and Sciences Press, 1976), p. 211.

47. Mao Zedong, "Speech to an Enlarged Politbureau Meeting," in *Long Live the Thought of Mao Tse-tung* (Taipei, 1969), p. 638.

48. Lieberthal, *A Research Guide,* p. 243.

49. Zhang Lifen, "The Limitations of the Law of Value in the Development of China's Grain Production," *Agricultural Economics* 10 (1981):30.

50. Lin Gang, "China's Agricultural Backwardness Is the Root Cause of its Rapid Population Growth," *Population Research* 1 (1981):22; Zhu Jingzhi, *China's Grain Policy and the Work to Supply Grain to Cities and Towns* (Beijing: Finance and Economics Publishing House, 1958), p. 17.

51. Lin Gang, "China's Agricultural Backwardness," p. 22.

52. Data on grain exported to international markets are taken from *Annual Economic Report of China 1982* (Hong Kong: Modern Culture Company, 1982), p. viii-47.

53. Nicholas R. Lardy, "Food Consumption in the People's Republic of China," in *The Chinese Agricultural Economy,* ed. Randolph Barker and Radha Sinha (Boulder, Colo.: Westview Press, 1982), pp. 147–62.

54. Nicholas R. Lardy, *Agriculture in China's Modern Economic Development* (Cambridge: Cambridge University Press, 1983), p. 87, table 2–8.

55. State Statistical Bureau, "A Collection of Statistical Materials on the Years 1949–1979," in *Annual Economic Report of China 1981,* ed. Xue Muqiao, p. vi-10.

56. Yang Jianbai and Li Xuezeng, "The Structure of Agriculture, Light Industry, and Heavy Industry," in *Research on Issues in China's Economic Structure,* ed. Ma Hong and Sun Shangqing (Beijing: People's Publishing House, 1981), p. 106.

57. For documentation and more detailed discussion of the developments in Fujian discussed below see Lardy, *Agriculture in China's Modern Economic Development,* ch. 2.

58. Tax, quota, and overquota deliveries actually were stable or fell slightly while negotiated-price deliveries rose. See ibid.

59. For documentation and more detailed analysis of state procurement and market prices in the late 1970s and early 1980s see Nicholas R. Lardy, "Agricultural Prices in China," World Bank Staff Working Paper No. 606 (Washington, D.C.: International Bank for Reconstruction and Development, 1983).

60. State Council, "Approval and Dissemination of a Report by the All-China Federation of Supply and Marketing Cooperatives Concerning Several Current Issues in the Purchase of Agricultural and Subsidiary Products," *New China Monthly* 8 (1981):137–39.

61. Mao Xincui, "A Great Advance in Periodic Market Trade," in *China's 1981 Encyclopedic Yearbook,* ed. Hu Qiaomu (Beijing and Shanghai: Chinese Encyclopedia Publishing House, 1981), p. 220.

62. State Council, "Order on Strengthening Market Management, Upholding Socialist Economics," *People's Daily,* January 16, 1981, p. 1.

63. State Council, "Notice on Firmly Stabilizing Market Prices," *New China Monthly* 1 (1982):130–31; Zhang Qi, "Guaranteeing Basic Stability of Market Prices Is the Number One Requirement of Present Price Policy," *Price Theory and Practice* 1 (1982):1–4.

64. Hu Yaobang, "Create a New Situation in All Fields of Socialist Modernization," *Beijing Review*, September 13, 1982, p. 18.

65. Lardy, "Agricultural Prices in China," table 3.3.

66. State Council, "Order Fixing Grain Procurement, Sales, and Transfers," *China Finance and Trade Journal*, March 20, 1981, p. 1; Ministry of Food, Rural Procurement and Sales Bureau, "Purchased Grain May Not Be Used to Fulfill Procurement Tasks," *Rural Work Bulletin* 3 (1982):45.

# 3

# Poverty and Progress in the Huang and Huai River Basins

## Thomas B. Wiens

The Huang and Huai River basins encompass an area of the North China plain including parts of the provinces of Hebei, Henan, Shandong, Anhui, and Jiangsu. The area is subject to erratic precipitation and overall climate. Most of the cultivated land is low-lying and poorly drained. Some 6 million hectares are saline. The area continues to be subject to both drought and flood. And yet in recent years selected districts, such as those along the Yellow River in Henan or south of Jinan in Shandong, have achieved high productivity and reasonable security through water conservancy measures. Two bases for commercial grain production lie in these basins, one in northwestern Shandong and the other in northern Anhui and Jiangsu. Both have low crop yields and the potential for more rapid improvement. In general, the contrast between prevalent poverty and isolated progress attracts attention to this area as one that could play a significant role in future national agricultural growth. I wish to consider below the means by which government policy and practice could enhance this role.

Even in the poorest parts of the area, instances of successful development may be found. A useful starting point for this discussion would be a comparison of such examples with their opposites. The statistics utilized pertain to eight production teams, selected in pairs by county. These were chosen in order to contrast the condition of teams that have undergone a state-assisted development program with "unimproved" teams that have not. The counties studied are typical of northwestern Shandong, northern Anhui, and eastern Henan. Following this comparison, the major instruments of government agricultural policy are considered. These include modification of farm price structure and level, reform of the institutional structure of farm management, and upgrading of farm support services. Since 1978, the central government has turned its attention to these instruments in succession, and some initial judgments can be made about the measures implemented. Beyond this, I will advance some opinions as to what further measures may be taken to spur more rapid agricultural development.

## The Anatomy of Poverty and Progress

A sample of production teams reflecting extremes of poverty and progress in the North China plain is statistically profiled in tables 3.1– 3.7. Of the eight teams surveyed, the four labelled ''A'' (collectively referred to below as Team A) are most representative of their counties, although the Qihe and Ningling teams A are actually worse off than average. To recall an oft-cited statement, ''Based on statistics, the annual foodgrain ration of 200 million peasants in China is less than 300 jin, that is to say they are living in a state of semi-starvation.''[1] These two teams, with per capita distributed grain rations of 250 and 270 jin respectively, are part of this group. However, as team members purchased (at official retail prices) enough wheat and corn to raise the grain consumption per capita to nearly 340 jin, the appropriateness of the term ''semi-starvation'' is questionable. Still, Team A is handicapped by saline soil and limited irrigation facilities; its crop yields in 1980 were in the vicinity of the national averages in 1957. It is at the starting point of the development spectrum.

Team B, on the other hand, is well along the development road by local standards. Two years or a decade earlier, its situation was comparable to that of Team A. Inclusion in a state-supported pilot project, not ''bootstrapping,'' has lifted it onto this road. To what combination of circumstances it owes this good fortune, it is fruitless to speculate. The important fact is that it has invested at least 50 yuan per mu in water conservancy and mechanization, including state grants, loans, and local cash and labor.

The contrasts between Team A and Team B are not always sharp. Team B may employ a slightly higher percentage of its labor force in cultivation, although aside from the ubiquitous brick factories, there are few alternatives in the collective sector (table 3.1). The amount of land per worker ranges from 3 to 9 mu, but it does not differ systematically between Team A and Team B. At first glance, there appear to be no clear differences in the proportion of acreage irrigated, but since a well that lacks a pump is still counted as irrigating a piece of land, the statistics here are meaningless.[2] Actually the reported costs of irrigation, electricity, and ''other'' (which includes diesel fuel) in table 3.4

indicate that Team B is applying much more water. Retained land ("private plots") as a percentage of total acreage varies from the extremes of the progressive Yucheng brigade, which allocates none, to the Ningling Team B, which has allocated the maximum currently allowed; in other cases the proportion continues to hover around 8 percent.

The differences in cropping patterns between teams A and B are subtle—the advanced team *may* put less emphasis on drought-resistant grain crops, more on cotton (a lucrative cash crop), and as part of its prescribed soil-improvement program, grows "green manure" crops (such as alfalfa), but there are exceptions to each generality. Team B does not necessarily multiple crop to a greater extent than Team A. Team B is clearly more mechanized, however, and the additional horse-power largely reflects engines or motors for driving tubewells.

Obviously Team B's crop yields are higher—especially wheat yields, where reliable irrigation makes the greatest difference (table 3.2). The effects of the differences in yields can be observed in table 3.3. The gross income of Team B is 70–400 percent higher than that of Team A. Moreover, Team A has very little cash income; most of its income comes in the form of distributed grain and other commodities. Most of its sales satisfy quota procurement requirements, and it sells little or nothing at above-quota prices (table 3.6). On the other hand, because of the higher yields of Team B and greater production of nonfoodgrain crops (in which quota requirements are low or nonexistent), a high proportion of its sales are at above-quota prices, even though its quotas are higher than those of Team A. The higher grain yields also spill over into greater gross income from animal husbandry—more collective pigs slaughtered or shoats sold and at higher weights (the impact on private animal raising is equally dramatic), due to greater availability of feedgrain and crop by-products.[3]

In short, once a team can escape from a "subsistence orientation" through higher crop yields, its cash income can rise dramatically. The structure of procurement pricing reinforces this tendency, even though additional burdens, such as higher quotas or pressure to raise more collective pigs, offset the potential gains to some extent.

Of course, part of the additional gross revenues is plowed back or absorbed in the form of increased input use, which is responsible for the higher yields in the first place (table 3.4). The leap in the use of modern inputs—chemical fertilizers, pesticides, machinery, electricity, etc.—is striking. The costs of traditional inputs—seed, recycled straw, plow-animal feed, pig manure, etc.—do not increase so rapidly or

Table 3.1

## Basic Statistics for Teams (A = unimproved; B = improved)

| | Shandong | | Shandong | | Anhui | | Henan | |
| --- | --- | --- | --- | --- | --- | --- | --- | --- |
| | Yucheng A | Yucheng B | Qihe A | Qihe B | Suixi A | Suixi B | Ningling A | Ningling B |
| Population (persons) | 145 | 327 | 188 | 160 | 116 | 201 | 213 | 180 |
| Persons/mu cultivated | 0.3 | 0.3 | 0.5 | 0.4 | 0.3 | 0.4 | 0.4 | 0.7 |
| *Labor force (persons)* | 65 | 131 | 64 | 55 | 45 | 87 | 110 | 78 |
| Agricultural | 63 | 131 | 57 | 50 | 40 | 84 | 99 | 73 |
| Side activities | — | — | — | — | 1 | — | 4 | — |
| Industry | — | — | 5 | 5 | 4 | 3 | — | 3 |
| Other (professionals) | 2 | — | 2 | 1 | — | — | 7 | 2 |
| *Cultivated acreage (mu)* | 420 | 1,206 | 420 | 460 | 373 | 556 | 480 | 246 |
| Collective | 385 | 1,206 | 390 | 420 | 346 | 525 | 438 | 246 |
| Paddy field | — | — | 40 | 1 | — | — | — | — |
| Upland | 385 | 1,206 | 350 | 420 | 346 | 525 | 438 | 210 |
| Irrigated | 310 | 800 | — | — | — | — | 328 | 176 |
| Unirrigated | 75 | 406 | — | — | — | — | 110 | 34 |
| Retained land | 35 | 0 | 30 | 40 | 27 | 31 | 42 | 36 |

61

| Sown area (mu) | 638 | 1,100 | 540 | 660 | 552 | 815 | 598 | 296 |
|---|---|---|---|---|---|---|---|---|
| Crop: | | | | | | | | |
| Wheat (%) | 45.4 | 36.4 | 27.8 | 36.4 | 38.0 | 40.6 | 41.8 | 37.2 |
| Maize | 44.7 | 36.4 | 29.6 | 34.8 | 3.1 | 1.2 | 23.4 | 22.0 |
| Rice | — | — | 7.4 | — | — | — | — | — |
| Sweet potato | — | — | — | — | 26.3 | 20.9 | 6.7 | 5.1 |
| Kaoliang | — | 4.5 | 7.4 | — | 0.9 | 6.1 | 6.7 | 5.1 |
| Millet | — | 2.3 | 1.9 | 1.5 | 6.5 | — | 6.7 | 6.8 |
| Soybean | 7.4 | 2.3 | 7.4 | 1.5 | 8.3 | 14.7 | 10.0 | 20.3 |
| Peanut | — | — | — | — | 1.8 | 1.5 | 2.2 | — |
| Sesame | — | — | — | — | — | 7.4 | — | — |
| Cotton | — | 18.2 | 9.2 | 12.1 | 3.3 | 5.2 | 2.5 | 3.7 |
| Jute | — | — | 7.4 | — | 6.0 | — | — | — |
| Squash | 0.9 | — | — | — | — | — | — | — |
| Vegetables | — | — | 1.9 | 2.3 | — | — | — | — |
| Green manure | — | — | — | 11.4 | — | 3.8 | — | — |
| Unidentified | 1.6 | — | — | — | — | 1.3 | — | — |
| Machinery (hp) | 60 | 355 | 24 | 72 | 12 | 56 | 22 | 30 |
| Indicators of intensity: | | | | | | | | |
| Multiple cropping index (%) | 171 | 157 | 138 | 157 | 155 | 161 | 157 | 155 |
| Machinery per mu (hp) | 0.16 | 0.29 | 0.06 | 0.17 | 0.03 | 0.11 | 0.05 | 0.14 |
| Labor days/cult. mu | 40 | 35 | 21 | 30 | 33 | 33 | 22 | 13 |

Table 3.2

# 1980 Crop Yields on Collective Sown Acreage (ton/ha)

| Crop | Shandong | | | | Anhui | | Henan | |
| --- | --- | --- | --- | --- | --- | --- | --- | --- |
| | Yucheng A | Yucheng B | Qihe A | Qihe B | Suixi A | Suixi B | Ningling A | Ningling B |
| Wheat | 1.215 | 3.788 | 0.606* | 2.280 | 1.222 | 2.280 | 0.720 | 3.075 |
| Maize | 1.995 | 4.058 | 1.918 | 3.660 | 3.320 | 1.875 | 1.095 | 2.190 |
| Sweet potato | — | — | — | — | 2.738 | 4.408 | 1.406 | 3.998 |
| Sorghum | — | 2.280 | 0.372 | — | 0.850 | 1.875 | 0.188 | 0.998 |
| Millet | — | 0.938 | 0.372 | 2.565 | 1.641 | — | 0.563 | 3.000 |
| Soybean | 1.125 | 0.923 | 0.667 | 1.148 | 0.769 | 0.938 | 0.600 | 1.050 |
| Green beans | — | — | — | — | 0.378 | 0.525 | — | — |
| Peanut | — | — | — | — | 0.450 | 1.425 | 1.043 | — |
| Sesame | — | — | — | — | — | 0.323 | — | — |
| Cotton | — | 0.827 | 0.113 | 0.840 | 0.299 | 0.488 | 0.375 | 0.109 |
| Jute | — | — | 0.300 | — | 0.265 | — | — | — |
| Squash | 3.000 | — | — | — | — | — | — | — |

*Yields on retained land (2 ha) were 1.0 tons/ha.

else decline, so that the percentage difference in total crop production costs between Team A and Team B is about the same order of magnitude as that of gross revenues. The law of diminishing returns does not play such a prominent role between these two levels of farming, because the weight of fast-growing modern inputs is not yet large. As Team B attempts to intensify further its production techniques, marginal costs may approach marginal revenues, simply because the weight in total costs of traditional inputs, which grow slowly if at all, is much reduced. That is, all we see here is a reflection of some general laws of agricultural development.

The growth rate of net income between Team A and Team B may be a bit higher or lower than that of gross revenues (table 3.5); the distributed income in turn grows somewhat less rapidly than total net income. Team A cannot set aside enough cash or commodities to provide more than a minimal amount of working capital (the ''production fund''), much less grain reserves or funds to cover depreciation of existing equipment. It lives from hand to mouth, and it has virtually no cash to cover operating costs in advance of the next harvest. its team members receive no cash at all as distributed income. Team B is better off in these respects, especially its cash distribution, but its rate of accumulation is not necessarily very high (the Yucheng Team B is a notable exception). Comparing the size of the production fund with costs of production, it is obvious that both teams require credit to cover operating costs.

The question that remains to be addressed is how does a typical ''Team A'' become transformed into a ''Team B''? The answer is, with more than a little help from the state. The essential difference between Team A and Team B is in extent and security of irrigation. In this region, the amount of surface irrigation is quite limited; indeed, the rivers and canals dry up in times of drought. Powered tubewells provide the only reliable source of water. Most of the teams in this sample, however, have or had serious salinity problems, which are compounded by irrigation unaccompanied by extensive drainage works. Team A can, as in Ningling, construct its drainage system with labor alone, although unless irrigation facilities are soon constructed, the motivation for improved drainage will not be strong. But the next step requires capital—about 270 yuan per well to hire a drilling rig and specialists, plus team labor. To equip a well requires at least 1,000 yuan if electricity is available, or 1,500 yuan for diesel power. And normally the Credit Cooperative will be reluctant to lend to Team A for more than working capital requirements, in view of its inability to repay larger loans. If it can scrape together 50–70 percent of the cash costs on its

Table 3.3

## Composition of Gross Value of Collective Income, 1980
(yuan/collectively cultivated mu)

| | Shandong | | Shandong | | Anhui | | Henan | |
|---|---|---|---|---|---|---|---|---|
| | Yucheng A | Yucheng B | Qihe A | Qihe B | Suixi A | Suixi B | Ningling A | Ningling B |
| *Collective revenue* | 46.81 | 115.40 | 22.89 | 114.58 | 46.74 | 80.37 | 27.47 | 90.65 |
| of which: | | | | | | | | |
| *Cultivation* | 37.79 | 110.49 | 22.89 | 101.35 | 42.87 | 75.67 | N/A | N/A |
| *Cash income* | 2.87 | 81.80 | 5.48 | 65.23 | 13.92 | 37.70 | — | 18.86 |
| Grain procurement | 2.21 | 3.48 | — | 17.85 | 5.38 | 7.42 | — | 0.59 |
| Above-quota grain | — | — | — | 5.71 | 4.20 | 8.54 | — | 5.41 |
| Addition to reserves | — | 2.57 | — | — | — | 3.09 | — | — |
| Cotton procurement | — | — | 3.85 | 28.57 | 2.03 | 3.33 | N/A | N/A |
| Above-quota cotton | — | 44.78 | — | 11.90 | 0.26 | 8.45 | N/A | N/A |
| Oilseed quota | — | — | — | — | 0.35 | 0.49 | N/A | N/A |
| Above-quota oilseed | — | 5.14 | 1.74 | — | — | 1.50 | N/A | N/A |
| Other procurement | — | — | — | — | — | — | — | — |
| Straw/firewood | 0.66 | 4.15 | — | 0.71 | 1.70 | — | N/A | N/A |
| Fruit | — | 6.21 | — | — | — | — | — | — |
| Feed to PC members | — | 3.93 | — | — | — | — | — | — |
| Other | — | 11.52 | — | 0.48 | — | 4.88 | N/A | N/A |

| | | | | | | | | |
|---|---|---|---|---|---|---|---|---|
| *Commodity income* | 18.73 | 28.68 | 17.30 | 36.11 | 28.95 | 37.97 | N/A | N/A |
| Distributed grain | 13.93 | 17.12 | 15.13 | 24.76 | 16.25 | 22.71 | 14.84 | 8.94 |
| Distributed straw | 2.79 | — | 1.21 | 2.86 | 0.87 | 1.35 | 0.26 | 2.09 |
| Distributed vegetables | 1.23 | 1.95 | 0.96 | 3.10 | — | — | 2.47 | 6.10 |
| Distributed veg. oil | 0.78 | — | — | 0.94 | 0.58 | 0.38 | — | 0.41 |
| Distributed cotton | — | 1.33 | — | 1.03 | 0.74 | 1.15 | 0.37 | — |
| Straw recycled | — | 2.61 | — | 3.43 | — | — | — | — |
| Other (seed, feed) | — | 5.68 | — | — | 10.51 | 12.38 | N/A | N/A |
| *Animal products* | 0.96 | 2.07 | — | 0.99 | — | 0.84 | N/A | N/A |
| Pigs sold | 0.25 | 0.34 | — | 0.52 | — | — | N/A | N/A |
| Pigs slaughtered | 0.52 | 1.07 | — | — | — | — | — | — |
| Shoats sold internally | 0.18 | 0.26 | — | 0.48 | — | — | N/A | N/A |
| Shoats sold on market | — | 0.40 | — | — | — | — | N/A | N/A |
| Sale of cattle | — | — | — | — | — | 0.84 | N/A | N/A |
| *Fisheries* | — | — | — | — | — | 0.14 | | |
| *Side activities* | 1.04 | 0.70 | — | 0.34 | 2.26 | 0.25 | N/A | N/A |
| Regular labor income | — | 0.20 | — | — | 1.54 | 0.11 | N/A | N/A |
| PC and brigade industry | — | 0.12 | — | 0.34 | 1.54 | 0.11 | N/A | N/A |
| Outside PC | — | 0.08 | — | — | — | — | N/A | 1.63 |
| *Temporary labor* | 1.04 | 0.50 | — | — | 0.72 | 0.13 | N/A | N/A |
| Brigade industry | — | — | — | — | 0.72 | 0.13 | N/A | N/A |
| Transport work | 1.04 | — | — | — | — | — | N/A | N/A |
| *Other* | 5.01 | 2.14 | — | 11.89 | 1.61 | 3.48 | N/A | N/A |

Table 3.4

**Composition of Collective Production Costs, 1980
(yuan/collective cultivated mu)**

| | Shandong | | Shandong | | Anhui | | Henan | |
| | Yucheng A | Yucheng B | Qihe A | Qihe B | Suixi A | Suixi B | Ningling A | Ningling B |
|---|---|---|---|---|---|---|---|---|
| *Total costs* | 19.70 | 52.08 | 8.64 | 59.88 | 27.56 | 40.86 | 13.86 | 34.13 |
| of which: | | | | | | | | |
| *Cultivation* | 16.16 | 35.92 | 8.26 | 44.95 | 15.34 | 31.13 | 9.68 | 25.61 |
| of which: | | | | | | | | |
| Seeds | 2.62 | 1.66 | 2.50 | 3.42 | 5.59 | 7.69 | 2.97 | 3.25 |
| Chemical fertilizers | 3.65 | 15.51 | 2.65 | 18.62 | 2.90 | 7.38 | 2.28 | 12.20 |
| Beancakes | 2.18 | — | — | — | — | — | — | — |
| Other chemicals | 0.03 | 1.00 | 0.13 | 1.90 | 0.07 | 0.95 | 0.23 | 0.41 |
| Recycled straw | 2.60 | — | — | 3.57 | — | — | — | — |
| Pig manure (purch.) | 0.54 | NC | 1.50 | 4.76 | 2.71 | 3.40 | — | — |
| Plow animal feed | 0.66 | — | — | 0.95 | — | 4.70 | 0.82 | 4.88 |
| Machine plowing | 0.91 | 4.81 | 0.64 | 0.95 | 0.03 | 0.77 | 0.68 | 0.41 |
| Other machine use | — | — | — | 2.38 | — | 0.19 | — | — |
| Electricity | — | 2.09 | — | 2.38 | 0.81 | 0.34 | — | — |
| Irrigation | 2.21 | 5.62 | 0.38 | — | — | 0.15 | 0.23 | 0.81 |
| Plastic sheet | — | — | — | — | 0.11 | 0.31 | 0.18 | 0.20 |

| | | | | | | | |
|---|---|---|---|---|---|---|---|
| Machine repair | 1.59 | 1.45 | 0.46 | 1.19 | 0.43 | 0.76 | 1.37 | 2.85 |
| Small tool repair | 0.70 | 1.09 | — | 0.71 | 1.53 | 0.67 | 0.46 | 0.61 |
| Other | 1.07 | 2.69 | — | 4.09 | 0.69 | — | 0.46 | — |
| Unused materials | — | — | — | — | 3.18 | — | — | — |
| *Animal products* of which: | 1.28 | 11.18 | 0.31 | 2.24 | 6.95 | 6.98 | — | — |
| Feedgrains | 0.65 | 3.73 | 0.30 | 1.19 | 4.23 | 4.70 | 1.64 | 4.88 |
| Other fine feed | — | 1.49 | — | — | — | — | — | — |
| Forage purchased | — | 5.59 | — | — | — | — | — | — |
| Feed processing fees | 0.05 | — | 0.01 | 0.48 | — | 0.38 | 0.18 | 0.16 |
| Other expenses | 0.58 | 0.35 | — | 0.57 | 2.72 | 1.90 | 0.07 | 2.03 |
| *Fisheries, etc.* | — | 0.25 | — | 0.27 | — | — | — | — |
| *Management* | 0.30 | 0.79 | 0.06 | 0.17 | 0.92 | 2.12 | N/A | N/A |
| *Taxes* | 1.07 | 0.95 | 0 | 0.56 | 2.49 | 3.03 | N/A | N/A |
| *Other* | 1.95 | 3.00* | — | 11.68** | 1.86 | 1.41 | N/A | N/A |

*Including depreciation.
**Losses on vegetables, subsidies for errands, entertainment. NC not counted in costs, because paid for in workpoints.

Table 3.5

## Distribution of Net Income, 1980
### (yuan/collective cultivated mu)

| | Shandong | | Shandong | | Anhui | | Henan | |
| | Yucheng A | Yucheng B | Qihe A | Qihe B | Suixi A | Suixi B | Ningling A | Ningling B |
|---|---|---|---|---|---|---|---|---|
| Net income | 27.10 | 66.39 | 14.24 | 54.70 | 19.18 | 42.91 | 22.08 | 66.43 |
| Distributed income | 21.00 | 39.80 | 14.22 | 46.96 | 18.01 | 37.67 | 17.58 | 61.66 |
| of which: | | | | | | | | |
| Cash | — | 20.73 | — | 16.72 | 0.03 | 12.09 | — | 19.05 |
| Commodities | 21.00 | 19.07 | 14.22 | 30.24 | 17.99 | 25.59 | 17.58 | 42.62 |
| Accumulation | 5.02 | 26.59 | 0.02 | 7.74 | 1.15 | 5.24 | 4.50 | 4.77 |
| of which: | | | | | | | | |
| Grain reserves | — | 4.65 | — | — | — | — | — | — |
| Depreciation fund | — | 1.16 | — | — | — | — | — | — |
| Production fund | 1.28 | 19.57 | — | 4.76 | 0.70 | 5.24 | 2.06 | 4.77 |
| Other | 3.75 | 1.21 | 0.02 | 2.98 | 0.45 | — | 2.44 | — |
| of which: | | | | | | | | |
| Welfare fund | 1.41 | — | — | — | — | — | — | — |
| Public accumulation | 2.35 | — | — | — | — | — | — | — |

Table 3.6

## Distribution of Grain and Other Products, 1980
### (jin of unprocessed product/collective cultivated mu)

| | Shandong | | Shandong | | Anhui | | Henan | |
| | Yucheng A | Yucheng B | Qihe A | Qihe B | Suixi A | Suixi B | Ningling A | Ningling B |
|---|---|---|---|---|---|---|---|---|
| *Total grain distributed* | 360 | 404 | 167 | 67 | 10 | 62 | 39 | 11 |
| of which: | | | | | | | | |
| 1979 residual | — | 26 | 16 | 115 | — | 8 | N/A | N/A |
| 1980 production | 360 | 378 | 151 | 453 | 310 | 454 | N/A | N/A |
| *Sold to state* | 13 | 26 | — | 118 | 52 | 107 | — | 43 |
| of which: | | | | | | | | |
| Quota sales | 13 | 26 | — | 36 | 38 | 60 | — | 9 |
| Above-quota sales | — | — | — | 82 | 14 | 46 | — | 34 |
| *Rations* | 166 | 182 | 121 | 250 | 172 | 252 | 102 | 306 |
| *Seed* | 22 | 17 | 25 | 59 | 38 | 52 | 18 | 29 |
| *Feed* | 12 | 76 | 4 | 82 | 44 | 44 | 16 | 19 |
| of which: | | | | | | | | |
| Collective animals | 6 | 27 | 4 | 39 | 44 | 44 | N/A | N/A |
| Private animals | 6 | 49 | — | 42 | — | — | N/A | N/A |
| *Reserves* | — | 19 | — | — | — | — | — | — |

Table 3.6 (continued)

**Distribution of Grain and Other Products, 1980**
**(jin of unprocessed product/collective cultivated mu)**

| | Shandong | | Shandong | | Anhui | | Henan | |
|---|---|---|---|---|---|---|---|---|
| | Yucheng A | Yucheng B | Qihe A | Qihe B | Suixi A | Suixi B | Ningling A | Ningling B |
| *Production grain** * | 147 | 81 | 18 | 59 | 4 | — | 2 | 10 |
| *Other crops* | | | | | | | | |
| *Seed cotton* | | | | | | | | |
| Sold to state | N/A | 52 | 2 | 26 | 2 | 29 | N/A | N/A |
| Quota sales | N/A | 51 | 2 | 25 | 1 | 18 | N/A | N/A |
| Above-quota sales | N/A | N/A | — | 3 | 1 | 6 | N/A | N/A |
| Distributed | N/A | 2 | — | 23 | * | 12 | 1 | * |
| *Oilseeds* | | | | | | | | |
| Sold to state | N/A | N/A | N/A | N/A | 5 | 5 | N/A | N/A |
| Quota sales | N/A | N/A | N/A | N/A | 1 | 3 | N/A | N/A |
| Above-quota sales | N/A | N/A | N/A | N/A | 1 | 1 | N/A | N/A |
| Market sales | — | — | — | — | 2 | — | — | — |
| Distributed | N/A | N/A | N/A | N/A | 1 | 1 | 4 | N/A |

*Less than 0.5.
**Grain used to provide rations for laborers on county water works, tractor drivers, cadres on official business, etc.

own, it may be able to borrow the remainder. Ningling's Team A was lucky enough to borrow all the capital required for drilling six wells; unfortunately, it was able to acquire only one pump, for lack of further capital, so most of its wells are inoperable. Ningling's Team B, on the other hand, was able by 1979 to invest 2,300 yuan of its own money and borrow 1,800 yuan to add three completely equipped wells. Most of the B teams surveyed, however, started with an injection of a grant of state capital, because of their fortuitous inclusion in a state pilot project, and were continuing to get special treatment.

In summary, the path of development from Team A to Team B in much of this area is rather obvious, and bootstrapping will not suffice. Cash investment of about 500 yuan per farm family is required, provided that ample shallow underground water supplies are available (a condition not satisfied in every locality). The rate of return on this investment is quite adequate by any measure for successful instances such as Team B. But neither local savings and long-term credit nor the resources of local governments are currently sufficient to provide the seed money for many teams.

Team B, however, may have been the exception rather than the rule even in the best of government pilot projects. Statistics for all the teams included in certain projects indicate that on average the rate of progress has been spotty and not much greater than that of teams outside the projects. The apparent reason is that coordination, management, and incentives have been weak or inconsistent. In concrete terms, improved irrigation was not accompanied by completed drainage systems or

Table 3.7

**Nitrogenous Fertilizer Application Levels
(kg/ha nutrient weight)**

|  | Shandong | | Shandong | | Anhui | |
|  | Yucheng A | Yucheng B | Qihe A | Qihe B | Suixi A | Suixi B |
|---|---|---|---|---|---|---|
| Wheat | — | 105 | 9 | 156 | 32 | 68 |
| Maize | 32 | 90 | 48 | 62 | 88 | 48 |
| Sweet potato | — | — | — | — | 21 | 68 |
| Kaoliang | — | — | — | — | — | 45 |
| Millet | — | — | — | — | 125 | — |
| Peanut | — | — | — | — | — | 45 |
| Cotton | — | — | 23 | — | — | — |
| Jute | — | — | 23 | — | — | — |

sufficient increases in chemical fertilizer use; the fertilizer was concentrated on a few crops (see table 3.7), while other crops suffered. Wells were sunk too close together, or into saline aquifers. Project works were poorly maintained or supervised, so that wells were unequipped or inoperative and drainage ditches sealed up to store water. Repair and support services were inadequate. Indeed, the most striking yield increases (for cotton) were the result of price and procurement policy changes, rather than project investments. In brief, past experience indicates that investment without other reforms will not yield fruit.

## Prices and Incentives

The Chinese government in 1979 made major adjustments in the level and structure of farm prices, because of concern that the existing price structure provided little incentive to produce and also affected the choice of crops in undesired ways. Government officials expect to make no further major changes in the structure of crop prices in the foreseeable future, although input prices could conceivably be adjusted downward if the profitability of the producing industries increases. Since further use of the price instrument has thus been ruled out, the question to be dealt with here is whether the new farm price structure provides the appropriate set of incentives and provides farmers with an adequate livelihood.

The words "appropriate" and "adequate" call for finely tuned criteria. First, we will distinguish between the *level* and *structure* of farm prices (including both input and output prices): the former determines the average profitability in farming and therefore farm livelihood relative to other occupations; the latter determines the relative profitability of different crops as well as the marginal profitability of using added amounts of modern inputs such as chemical fertilizer.[4] It is possible for one of these to be "appropriate" or "adequate" and the other to be otherwise. Similarly, we will distinguish *intensive* from *extensive* farming: the former is high-yield farming with high operation costs, as generally practiced by "model" production teams or in areas of high multiple cropping rates in China; the latter is more typical of the North China plain. It is not only possible but highly probable that price level and structure that is adequate for the extensive farm is painfully inadequate for the intensive farm, because in China intensification has frequently been pushed far beyond an economically optimal level (under any reasonable price system). For the intensive farm, the most immediate remedy is deintensification rather than price reform. I will not deal here with its plight.[5]

Of course incentives are only meaningful if producers have sufficient freedom to act on them. If the incentives provided by the price *structure* are to be effective, the producers must be free to vary the acreage planted in each crop and the amounts of each major productive input employed. If the incentives provided by the price *level* are to be effective, there must be some occupational mobility into and out of farming or at least some freedom to vary the managerial effort and time invested in farming. Since we are concerned primarily with collective farming, the latter includes changes in the proportion of effort devoted to private plots and sideline activities versus the proportion invested in collective farming. For the moment, I will assume the existence of such freedoms; later I will reconsider this assumption.

The current farm price structure encompasses three distinct price levels. Quota prices apply to crops sold in fulfillment of procurement quotas; above-quota prices to crops sold in excess of the quota. These prices are basically fixed and uniform for equivalent quality produce, regardless of differences in transport cost. Above-quota prices are exactly 50 percent higher than quota prices for grains and oilseeds, approximately 30 percent higher for cotton, and 20 percent for baste fibres.[6] Norms or obligations for above-quota sales are determined one or two months before each harvest, based on projected yields. Negotiated prices (*yijia*) fluctuate at approximate parity with free-market prices. They apply to sales by individual farmers or (rarely) teams to the state that could otherwise legally be sold on the free market. In the North China plain in 1980, negotiated prices were usually 10–20 percent higher than above-quota prices. However, in at least some provinces, above-quota sales also entitle a team to purchase special allocations of chemical fertilizer. For example, in Shandong one ton of ammonium-sulfate equivalent is allocated per ton of above-quota grain; the ratio is 2:1 for cotton, soybeans, and oilseeds. This bonus does not apply to sales at negotiated prices.

Quotas for most products were last fixed in this area in 1971, technically for a period of five years, but in fact they have remained unchanged since. They are fixed for two major product categories—grain and cotton—and apparently apply to the latter only if it was produced in 1971 by a particular team. If no quota exists for a product, the above-quota price normally applies. Starting in 1981, cotton may be substituted for grain in quota fulfillment at a ratio of 1:2, but only if the grain quota would otherwise remain unfulfilled. Quotas for other crops, such as peanuts, vary from year to year. Local officials expect quota levels to continue unchanged even if productivity increases rapidly; however, they could be adjusted at the option of Beijing. Based on incomplete data, quota procurements in the areas studied represent about 10 percent of grain production (but several areas have been partially or wholly exempted from quota procurements because of low production levels); 20–40 percent of oilseeds; and 50–80 percent of fiber crops. Sales at above-quota prices as a proportion of total production range from 3 to 12 percent for grain, 0 to 22 percent for oilseeds, and 0 to 29 percent for fibers. Proportions retained or distributed to team members represent about 75 to 92 percent of grain (100 percent where procurement exemption has occurred), about 50 percent of oilseeds, and 30 percent of cotton.

Since such a large proportion of production is retained or distributed, we must consider how this portion should be valued in monetary terms. Present accounting conventions in China use "internal prices" (actually 1978 quota prices) to evaluate such commodity income, lest the 1979 price reforms appear to have increased total farm incomes more than was actually the case. But free-market or, equivalently, negotiated prices are more appropriate for our purposes, as they reflect rural consumer demand and also the amount the farmer can obtain if he sells part of the commodities he receives from the production team. Consequently, in computing average "domestic prices" received at the farm level, I have used negotiated prices to value the proportion of crops not sold at quota or above-quota prices (and similarly, the seed and feed expended in production).

Now, in examining the incentive effect of the price *level*, we are interested in the average value of production, which is a weighted average of quota and above-quota sales prices and the negotiated price; the weights are the amounts sold at the first two prices and the amounts retained or distributed respectively. In examining price *structure*, the *marginal* value of production receives our attention, since it is the comparison of marginal values for different crops that should induce changes in allocation of land among crops, and it is the comparison of marginal value of additional output with marginal costs of required additional inputs that induces changes in input use. Two cases can be distinguished: for teams that cannot meet their quota sales requirements (e.g., Team A above), the marginal price received is the quota procurement price. For the majority of teams, which do meet quota requirements, most of any additional production will be sold at above-quota prices and the remainder will be retained, so their marginal product is priced at or above the above-quota procurement price.

The final, and most difficult, step is to determine standards of adequacy against which China's farmgate price structure and price levels can be compared. Because China is a significant net importer of most crops as well as agricultural chemicals, the domestic price *structure* is best appraised relative to the border prices[7] of the same crops. In fact, the price structure that maximizes the supply of agricultural products that can be obtained from a combination of domestic production and trade is one in which the relative marginal prices of crops are the same as the corresponding border price relatives. Moreover, unless there are systematic differences in Chinese agricultural production functions from those of other producer countries, the domestic ratios of product to input prices should also be the same as those of border prices. For additional production of a particular crop enables the state to reduce imports or increase exports of that crop (or close substitutes), at the expense of increased imports of the inputs in production, e.g., fertilizers and agrochemicals. Domestic price ratios differing from those of border prices would provide incentives for expansion or contraction of production of some or all crops that would entail a net loss of foreign exchange.

Estimates of 1980 border prices, expressed as farmgate prices, are derived in table 3.8, assuming all crops in the table are imported (actually, rice and peanuts are exported, and rapeseed and cottonseed are not traded; similar computations for rice treated as an export good are given in the notes to table 3.9). Table 3.9 compares these with average domestic prices, negotiated, above-quota, and quota prices. These are also expressed with the price of wheat or urea as a numeraire. It may be observed that the 1979 price reforms brought about a price *structure* that meets our criterion of adequacy, at least for major traded crops and inputs. With the price of nitrogenous fertilizer as numeraire, the price relatives for wheat, maize, soy, and cotton based on border prices are 0.85, 0.62, 1.17, and 8.07 respectively. Taking above-quota prices at the closest approximation to a marginal domestic price, the corresponding relatives are almost identical: 0.90, 0.65, 1.23, and 8.14 respectively. At first glance, this identity does not seem to hold true for rice, but closer examination shows that it does. The use of a border price based on import cost is inappropriate, as China is a net exporter of rice; based on the 1979 average export price, the border price adjusted to the farmgate level would be 334 yuan per ton, and the price relative with urea as numeraire would be 0.70. Also, North China

Table 3.8

## Derivation of Farmgate Prices of Crops and Farm Inputs From Border Prices, 1980 ($/ton or yuan/ton)

| 1980 price | Wheat (a) | Maize (b) | Rice (c) | Soy (d) | Peanut (e) | Rape (f) | Cotton seed (g) | Cotton (h) | DAP (i) | Urea (j) | TSP (k) |
|---|---|---|---|---|---|---|---|---|---|---|---|
| 1 Import price $ | 191 | 125 | 315 | 296 | 493 | 326 | 246 | 2,070 | 222 | 222 | 180 |
| 2 Ocean freight $ | 36 | 36 | 36 | 19 | 19 | 19 | 19 | 19 | 36 | 36 | 36 |
| 3 CIF Qingdao $ | 227 | 161 | 351 | 315 | 512 | 345 | 265 | 2,089 | 258 | 258 | 216 |
| 4 CIF Qingdao Y | 386 | 274 | 597 | 536 | 870 | 587 | 451 | 3,551 | 439 | 439 | 367 |
| *Plus:* | | | | | | | | | | | |
| 5 Port charges | 25 | 25 | 25 | 25 | 25 | 25 | 25 | 25 | 25 | 25 | 25 |
| 6 Transport: port-wholesalers | 5 | 5 | 5 | 5 | 5 | 5 | 5 | 3 | 3 | 3 | 3 |
| *Minus:* | | | | | | | | | | | |
| 7 Transport: farm-wholesalers | -10 | -10 | -10 | -10 | -10 | -10 | -10 | -10 | +10 | +10 | +10 |
| *Equals:* | | | | | | | | | | | |
| 8 Price ex mill | 406 | 294 | 617 | 556 | 890 | 607 | 471 | 3,571 | 477 | 477 | 405 |
| *Times:* | | | | | | | | | | | |
| 9 Adjustment | | | .70 | | .65 | | | | | | |

| | | | | | | | | | | | |
|---|---|---|---|---|---|---|---|---|---|---|---|
| *Minus:* | | | | | | | | | | | |
| 10 Milling cost | — | — | 29 | — | 24 | — | — | 130 | — | — | — |
| *Plus:* | | | | | | | | | | | |
| 11 Byproducts | — | — | 9 | — | 24 | — | — | 292 | — | — | — |
| *Equals:* | | | | | | | | | | | |
| 12 Farmgate price | 406 | 294 | 412 | 556 | 579 | 607 | 471 | 3,850 | 477 | 477 | 405 |

a U.S. No. 1 Soft Red Winter Wheat, FOB, Gulf ports.

b U.S. No. 2 Yellow, FOB, Gulf.

c Thai 5% broken, FOB, Bangkok, adjusted for quality assuming rice milling standards: 10% high quality (5% brokens); 60% medium grade (25-30% brokens); 30% lower quality (42% brokens). Conversion factor applied to high-quality rice is 0.725.

d-g CIF, European ports.

h Mexican lint SM1-16" CIF, N. Europe.

i Diammonium phosphate, bulk, FOB, Florida.

j Urea, bagged, FOB, NW Europe.

k Triple Superphosphate, bulk, FOB, Florida.

2 Based on following data: average freight rate of wheat, Gulf-Japan, 1979-80 = $36/ton; Gulf-Rotterdam = $17/ton; difference = $19/ton.

6 Average rail freight charge from nearest port (Qingdao, Lienyungang, or Tianjin) to food-deficit municipality nearest county, based on distance and freight rate schedule.

7 Assuming 50 km between collection or distribution point and consumer/distributor location (nearest municipality and truck freight rates of Y0.20/ton-km for agricultural products, Y0.28/ton-km for inputs, and Y0.50/ton surcharge for trips over 25 km.

9 Adjustment from milled rice to paddy-equivalent price (70%); shelled to unshelled peanuts (65%).

11 For cotton, byproduct value is that of cotton seed, derived from border price on assumption that seed represents 62% of seed cotton weight.

Table 3.9

**Border and Domestic Prices and Price Relatives of Crops and Farm Inputs, North China, 1980 (yuan/ton)**

| 1980 price | Wheat | Maize | Rice (a) | Soy (b) | Peanut | Rape | Cotton seed | Cotton | DAP (c) | Urea (d) | TSP (d) |
|---|---|---|---|---|---|---|---|---|---|---|---|
| Border price | 406 | 294 | 412 | 556 | 579 | 607 | 471 | 3,850 | 477 | 477 | 405 |
| Domestic price | 557 | 363 | 380 | 784 | 1,026 | 984 | 223 | 3,730 | 542 | 553 | 429 |
| Negotiated price | 580 | 380 | 400 | 670 | 1,200 | 1,220 | — | 4,980 | — | — | — |
| Above-quota price | 500 | 360 | 340 | 680 | 1,020 | 1,020 | — | 4,500 | — | — | — |
| Quota price | 340 | 240 | 240 | 470 | 680 | 680 | — | 3,440 | — | — | — |
| *Wheat as numeraire:* | | | | | | | | | | | |
| Border price | 1.00 | 0.72 | 1.01 | 1.37 | 1.43 | 1.50 | 1.16 | 9.48 | 1.17 | 1.17 | 1.00 |
| Domestic price | 1.00 | 0.65 | 0.68 | 1.41 | 1.84 | 1.77 | 0.40 | 6.70 | 0.97 | 0.99 | 0.77 |
| Negotiated price | 1.00 | 0.66 | 0.69 | 1.16 | 2.07 | 2.10 | — | 8.59 | — | — | — |
| Above-quota price | 1.00 | 0.72 | 0.68 | 1.36 | 2.04 | 2.04 | — | 9.00 | — | — | — |
| Quota price | 1.00 | 0.72 | 0.68 | 1.36 | 2.04 | 2.04 | — | 10.12 | — | — | — |

*Urea as numeraire:*

| | | | | | | | | | | | |
|---|---|---|---|---|---|---|---|---|---|---|---|
| Border price | 0.85 | 0.62 | 0.86 | 1.17 | 1.21 | 1.27 | 0.99 | 8.07 | 1.00 | 1.00 | 0.85 |
| Domestic price | 1.01 | 0.66 | 0.69 | 1.42 | 1.86 | 1.78 | 0.40 | 6.75 | 0.98 | 1.00 | 0.78 |
| Negotiated price | 1.05 | 0.69 | 0.72 | 1.21 | 2.17 | 2.21 | — | 9.01 | — | 1.00 | — |
| Above-quota price | 0.90 | 0.65 | 0.61 | 1.23 | 1.84 | 1.84 | 1.84 | 8.14 | — | 1.00 | — |
| Quota price | 0.61 | 0.43 | 0.43 | 0.85 | 1.23 | 1.23 | — | 6.22 | — | 1.00 | — |

*Note*: All except border and domestic prices rounded to nearest Y10.

*Border price*—import prices, adjusted for transport, processing costs, and milling losses, to the equivalent of a farmgate price.

*Domestic price*—a weighted average of quota, above-quota, and negotiated prices, the weights being amounts sold at quota prices, above-quota prices, or retained respectively in the nine-county area. It is assumed that retained crops (including private sales) should be valued at free market prices or equivalently negotiated prices.

*Negotiated price*—prices which the government pays for grain sold by private individuals out of amounts distributed to them by production teams. Roughly equivalent to free market prices.

*Above-quota price*—prices paid for team sales in excess of quota procurement; 50% higher than quota prices for grain and oilseeds.

*Quota price*—prices paid for team sales in fulfillment of quota procurement obligations.

*Other notes*:

a  Border price, estimated as a farmgate price from the rice *export* price, is Y334/ton; with wheat as a numeraire, the price relative would be 0.82; with urea as numeraire, it would be 0.70.

b  The domestic price of soy is a *projected* one, assuming a 20% increase in quota price occurs shortly. Other prices are present actual prices.

c  Domestic price is the actual price of imported diammonium phosphate delivered at the farm level.

d  For urea (46% N) and TSP (46% $P_2O_5$), domestic price is average of local products (ammonium bicarbonate, ammonium sulphate, and urea; superphosphate) based on nutrient equivalence.

produces *Indica* rice varieties for self-consumption, whereas exports are of higher-quality *Japonica*. The above-quota price of *Japonica* in Jiangsu is 408 yuan per ton, and the price relative would be 0.74, which is very close to the relative for border prices.

By the same token, the domestic price structure does not provide sufficient incentive for use of inputs such as fertilizer to those teams that normally cannot meet their procurement quotas—presumably few in number. More commonly, teams that farm with excessive intensity (and therefore have input marginal productivities lower than those in other countries) may find that marginal costs exceed marginal revenues, even though their marginal sales are at above-quota prices.

As yet we have drawn no conclusions about the absolute price *level*, which has most bearing on how profitable farming is overall to the farmers. If all crop prices and the prices of major industrial inputs were raised (relative to the prices of consumer goods), it would not affect the price relatives discussed above, but it would increase absolute profitability. However, it is difficult to find an objective criterion against which to appraise the post-1979 farm price level. Procurement prices—even above-quota procurement prices—are surely lower than the level required to induce voluntary sales of an equivalent amount of produce. That is, they do involve a tax on the farm sector. The procurement quota, since it is fixed in absolute amount but varies among teams in proportion to land productivity and cultivated land per capita, cannot be faulted as a tax device (although the variable above-quota procurement norms can). That farmers would rather retain additional production than sell it to the state at procurement prices is not surprising, since thereby they evade the tax. On the other hand, the underpricing of procurements, which is largely passed on to consumers through low retail prices, enables the government to hold down the urban cost of living, and thus industrial wages, and to maintain high industrial profits, from which government revenues are ultimately derived. Any judgment concerning the appropriateness of the farm price level would depend on what is done with these government revenues. If they were directly or indirectly funneled back into agricultural investment, as opposed to a variety of productive and nonproductive alternative uses, surely our appraisal would not be unaffected. Since the appropriateness of the price level is not merely a matter of the adequacy of farm livelihood, I am unwilling to draw any conclusions about it here.

I do not believe that farmers are taking a loss on any major crop, except where excessively intensive production techniques have been adopted. Chinese production teams pay no fixed wages or rents and make no allowances for depreciation. Recent Chinese studies that discovered such "losses" computed output value at quota procurement prices, but attributed both fixed wages and depreciation as part of costs. At what wage rates could labor costs be evaluated? There have been three common methods, using the average value of the labor day, as computed from distributed income and work days recorded; the fixed wage rate paid on nearby state farms; or an accounting wage of 0.80 yuan a day, which had been an official national standard since the 1950s. Under the first standard, if all income was from cultivation and most was distributed, invariably some crops would appear to be money losers and others profitable. Under either of the remaining standards (the last was and is most common), many teams would appear to lose money on most crops prior to 1979. In the area studied here, that would still be true today, since the average nominal value of income distributed per workday remains around 0.50 yuan.

A more reasonable approximation of the return to cultivation is shown in table 3.10, based on northern Anhui data. The principles of valuation employed are as described earlier, so that gross and net value added per hectare are considerably higher than the teams' own accounts (using low internal accounting prices) or official accounting (using quota prices) would show. No attempt is made to price labor; rather, we present value added per hectare, value added per workday, and value added per yuan of working capital as rough indices of economic returns (each of which includes what in nonsocialist agriculture would be called rent, wages, and profit).

Labor is not priced for several reasons. First, the accounting "workday" here is not a measure of time and effort, but simply equals ten workpoints. This is probably equivalent to only 50 to 70 percent of the time and effort of an adult male worker engaged, for example, in water conservancy construction work. The cash value of a ten-workpoint workday in this area is about 1 yuan (pricing commodities distributed at their market value). However, this is not a fixed wage, but the value of income distributed; it may include implicit rental return and profit. It is not, in any case, the opportunity cost of labor: I estimate that crop cultivation, animal husbandry, and forestry need absorb only about one-third of the rural labor force (more than one-half at peak periods). There is therefore considerable underemployment. In this circumstance, labor's opportunity cost is very low, and it should be virtually disregarded in appraising relative or absolute profitability in farming.

Table 3.10 should be interpreted with caution, particularly since

Table 3.10 **Estimated Return to Cultivation by Crop, Suixi County, Anhui (a) (yuan/ha)**

| | Wheat | Rice | Soybean | Sweet potato | Maize | Millet | Kaoliang | Cotton | Sesame | Rape-seed | Peanuts | Average |
|---|---|---|---|---|---|---|---|---|---|---|---|---|
| Net value added | 283.97 | 1,419.08 | 701.91 | 784.55 | 886.61 | 501.11 | 747.21 | 454.61 | 306.70 | 244.88 | 834.89 | 519.32 |
| Total revenues (b) | 653.16 | 1,995.71 | 867.24 | 975.00 | 1,097.17 | 654.64 | 890.52 | 868.04 | 472.58 | 371.17 | 1,247.72 | 795.62 |
| Total cost | 369.19 | 576.63 | 165.33 | 190.45 | 210.56 | 153.53 | 143.31 | 413.43 | 165.88 | 126.29 | 412.83 | 276.30 |
| Seed (b) | 119.70 | 118.20 | 49.31 | 55.58 | 22.28 | 5.75 | 5.73 | 29.10 | 13.50 | 13.73 | 270.00 | 79.35 |
| Fertilizer | 129.75 | 164.25 | 36.00 | 73.95 | 81.45 | 70.95 | 58.50 | 176.25 | 70.95 | 50.25 | 60.00 | 96.75 |
| Pesticide | 2.70 | 33.75 | 1.50 | 30.00 | 5.25 | 3.75 | 4.50 | 75.00 | — | — | 0.75 | 12.30 |
| Draft animal (b) | 49.54 | 96.93 | 38.77 | 12.92 | 43.08 | 43.08 | 43.08 | 43.08 | 43.08 | 40.93 | 43.08 | 38.56 |
| Tractor plowing | 9.00 | 15.00 | 1.50 | — | — | — | — | 9.00 | — | — | — | 4.65 |
| Irrigation | 4.50 | 67.50 | — | — | 21.00 | — | — | 3.00 | — | — | — | 2.55 |
| Other | 12.00 | 22.50 | 12.00 | 4.50 | 4.50 | 4.50 | 4.50 | 15.00 | 12.00 | 6.00 | 12.00 | 9.75 |
| Depreciation | 14.25 | 24.00 | 12.00 | 12.00 | 12.00 | 12.00 | 12.00 | 12.00 | 12.00 | 12.00 | 12.00 | 13.05 |
| Tool repair | 12.75 | 13.50 | 10.50 | — | 9.00 | 9.00 | 10.50 | 22.50 | 9.00 | 7.50 | 9.00 | 9.60 |
| Other repair (c) | 10.50 | 15.00 | — | — | 7.50 | — | — | 10.50 | — | — | — | 5.25 |
| Management (d) | 4.50 | 6.00 | 3.75 | 1.50 | 4.50 | 4.50 | 4.50 | 18.00 | 7.50 | 4.50 | 6.00 | 4.50 |
| Labor days | 173 | 437 | 81 | 196 | 259 | 161 | 144 | 621 | 121 | 104 | | |
| VA/Day | 1.64 | 3.25 | 8.67 | 4.00 | 3.42 | 3.11 | 5.19 | 0.73 | 2.54 | 2.35 | 3.02 | 2.78 |
| VA/Cost | 0.77 | 2.46 | 4.25 | 4.12 | 4.21 | 3.26 | 5.21 | 1.10 | 1.85 | 1.94 | 2.02 | 1.88 |
| Acreage % | 44 | 0 | 15 | 23 | 1 | 3 | 6 | 5 | 2 | 0 | 2 | — |

a A weighted average of data from three production teams, Suixi county, Anhui.

b Product and byproduct value valued in average domestic prices (weighted average of quota, above-quota, and negotiated prices). Cost of seed adjusted from valuation in quota prices to valuation in average domestic prices; draft animal feed costs similarly adjusted, based on ratio of average domestic price to quota price of maize.

c Including material costs in repair or improvement of land layout, irrigation and drainage systems, etc.

d Material expenditures and incentive payments to team leaders.

yields pertain to a single year, which may have been unusually good or bad for particular crops (cotton yields have since doubled, for example). Moreover, value added for each crop gives no clear guidelines on choice of crop, since the returns per hectare have to be appraised with respect to the resources available to a specific team, such as labor, working capital, and irrigation or drainage systems; to the variability of the returns (varying susceptibilities to weather vagaries and pests); and to the sequencing of labor and land requirements. For example, wheat and rape are the only winter crops, and wheat can be followed by another crop in the summer season, whereas all other crops absorb an entire crop year. Rice and maize generally require investments in irrigation systems, whereas the other spring-summer crops do not. The yields of summer crops are highly variable: differing degrees of risk have to be weighed against relative profitability, and the possibility of hedging against risks through diversification must be considered. Cultivation activities for crops sharing the same season have different timings, allowing diversification to reduce peak concentrations of labor demand. Perhaps the only firm conclusion to be drawn is that none of the crops can be rejected outright as involving losses to the team.

Aside from the effects of changes in the farm price level on income distribution, is there any effect on agricultural productivity? If so, it might be arguable that further increases in the farm price level would contribute to agricultural growth. The preliminary evidence for the nine counties is that such *general* incentive effects are minimal. Specifically, examination of eleven-year crop yield series for these counties does not suggest that the general price increase of 1979 of over 20 percent had any special effect on average crop yields in either 1979 or 1980. To be sure, wheat yields increased in one or both years, but over the whole period wheat yields have grown in tandem with improved irrigation and increased fertilizer supplies despite absence of price increases, and the 1979–80 yield increases were not out of line with this relationship. Yields of most other crops did not deviate from the normal range of fluctuation due to weather conditions, or even declined significantly. There are two exceptions: yields (and acreage) of both cotton and peanuts increased tremendously in 1980, reversing a long downward trend. It was explained locally that this was due to a combination of a change in policy designed to increase production of these crops, changes in the *relative* prices of these crops compared to grain, and application for the first time of above-quota pricing for these cash crops. These exceptions, if they reflect more than administrative fiat, seem to indicate that farmers would respond to price increases for one or a few crops but largely at the expense of other crops (e.g., of miscellaneous grains, the yields and area of which declined).

## Implementation of Rural Reforms

If farmers have no freedom to act on price signals, there is no way for the latter to lead the way to greater efficiency. This freedom unquestionably did not exist in China from the late 1950s. Therefore the significance of the reform of the agricultural price structure is open to question if it is not accompanied by organizational reforms that allow the production teams greater decision-making freedom. Rather dramatic organizational reforms have indeed occurred in the Huang and Huai River basins, and not just on paper. However, there is still some question as to whether the reforms will be enduring or go far enough.

I have written elsewhwere of the broad changes in rural policy beginning in 1978, including the attempt to eliminate "egalitarian" practices, "unreasonable burdens, and arbitrary exactions," and the use of coercion and regulation to control peasant farming practices.[8] As far as an outsider can observe, these reforms have begun to take hold in the area studied. Local officials seem committed to the improvement of farm incomes, and much more scope for private activity exists—private incomes now represent over 60 perent of total distributed income, almost all of this income from animal raising and private plots. In Henan and Anhui, officials talk of extending credit to encourage groups of individuals or single households to buy small tractors or sprinkler-irrigation equipment to perform custom work for their neighbors. Local development plans emphasize diversification out of grain and into oilseeds and cotton, animal husbandry, sericulture, orchards and trees, processing, and so on (provided that grain production is simultaneously increased).

The most dramatic changes occurred in 1981 as part of the implementation of various forms of the "responsibility system." In these poor counties, well over 90 percent of production teams adopted the most decentralized managerial form allowed, the "household responsibility system" (HRS), which involves complete land division with abdication of managerial responsibilities by the team.[9] This breakup of collective management occurred progressively but rapidly over the 1981 growing season. The teams remaining under less decentralized forms of management were those that had been more successful or beneficiaries of state largess.

Under the HRS, all cultivation activities for assigned pieces of land are the responsibility of a particular household. Where the team or other collective body supplies services, fees may be charged the household. Households are still responsible for meeting their share of quota and above-quota procurement obligations, the agricultural tax, feed for collectively owned animals, and contributions to funds for collective welfare and accumulation, if any. These responsibilities are enumerated in terms of summer- and autumn-harvested grain. The leadership of the production team has as its primary remaining responsibilities the collection of this grain as well as its function as compiler of farm-level statistics. Any other state-imposed obligations are presumbly still enforced by the team.

Despite the decentralized management, freedom to modify crop allocations is limited. Cotton and oilseed *acreage* is specified by the state, down to the team and household level. This is intended to increase production of these crops, but because the responsibilities are spread around, freedom to specialize is limited (although some areas are given higher acreage targets than others). Grain *quotas* are determined *by variety* down to the team and household level. However, with complete land division, the household can change the variety and purchase the required deliveries from the free market. In practice, this may not be common—if any use is to be made of collective machinery or irrigation facilities, it is advantageous for adjacent plots to share the same crops (and seed varieties).

Any institutional change introduces new problems, and this one is no exception. Some of the problems are undoubtedly temporary. For example, national regulations forbid the sale or rental of subdivided team land or the construction of houses or graves. But in a few areas visited, the last prohibition was not being observed—new grave mounds littered the fields in such a way as to be disruptive to cultivation. The irrationality of this situation is too obvious to endure for long.

A more permanent problem concerns the management of team assets other than land, especially machinery. Under HRS, some machinery has gone unutilized, including that which was previously unworkable or too large in scale, or because households wish to save on operating costs. For example, although many teams in the area own stationary power threshers, none was observed in operation during the wheat harvest. Instead, all households were using the public highways as threshing grounds, relying on passing vehicular traffic to thresh their individual piles of wheat. With peasants rushing to buy draft animals and few tractors observed in the field during the fall plowing season, local statistics on the proportion of land that is tractor plowed seemed greatly inflated.

Other machinery has been divided up among groups or households, which at least leads to better maintenance. But when the team rents its machinery to work groups or families, there is no guarantee that it will be maintained and returned in good condition. It has been suggested that an HRS should be devised for machinery management as well as field management.

The nature of the market for machinery of course has changed greatly. Due to the reduced role of the team and the more direct concern of team members for economy, there has been a sharp drop in demand for large-scale machinery. At the same time, demand for small-scale machinery has increased. The latter is still sold primarily to teams, secondarily to work groups, and rarely to individual households. But there have been cases of families buying tractors from teams and operating (illegal) custom transport businesses. Purchase of a tractor entitles one to a fixed, limited ration of fuel, but more can be obtained through the "back door." It is at the level of the work group that some local officials see a legitimate new market, for equipment to be used on their own fields and on a custom basis for other groups or households. At the same time, as the scale of production is reduced, it is expected that the demand for custom services based on more expensive equipment will rise. For example, Henan counties have established sprinkler-irrigation companies and are considering the operation of plant-protection companies, which for a fee would guarantee the fields of work groups or households against losses due to insects or disease.

The managerial situation has changed so drastically in such a short time that the administrative apparatus has not been able to catch up. Local cadres still think and talk in terms of an agriculture collectively managed by the production team, which continues to be their main point of contact with the grass roots. Yet is the team now anything more than a tax collecting body? How seriously could team leaders take their remaining responsibilities (such as the collection of statistics) when the attention of the members has turned exclusively toward enriching their families? Moreover, what proportion of local cadres regard this as a reversible "breathing spell" for the peasantry, as opposed to a permanent reform? Until local government adjusts to the reforms, these areas will be stuck in a situation akin to the early days of the cooperatives, when most of its resources and services were concentrated on the minority of farmers who chose to farm collectively.

It is too soon to tell whether the managerial reforms are having a significant effect on agricultrual productivity; Chinese assessments, attributing 1981 production growth to such effects, were undoubtedly exaggerated to justify the reforms. The safest prediction is that the number of man-hours expended in cultivation will be reduced with no reduction in work accomplished, and this labor wil be diverted to side activities, such as house building or commerce. Peasant parsimony will insure some reduction in costs of production, such as through the abandonment of unessential or costly machinery. And in North China there has already been considerable restoration of such cash crops as cotton and peanuts, due more to relative price increases and administrative fiat than to managerial reform. These all represent gains in efficiency and peasant incomes. Whether this will ultimately put more food on the table remains to be seen.

## Farm Support Services

In the realms of price and managerial reform, the Chinese leaders have already shot their bolts. If these reforms do not prove to be panaceas for the problems of the agricultural sector, then more attention will have to be given to the improvement of farm support services. At the very least, support services now geared to collective farming will have to be reoriented to serve the individual farm household. These services include agricultural research and extension and provision of seed, fertilizer, agrochemicals, irrigation, credit, machinery, and repair services. Taken in combination, strong and well-coordinated farm support services could result in substantial improvements in farm productivity. At present some crippling weaknesses in these services can be observed in the North China plain. The following is an abbreviated summary of the weaknesses evident to this observer.

### Agricultural Research and Extension

A loosely coordinated infrastructure ranging from central research organs such as the Chinese Academy of Agricultural Sciences (CAAS) down to the production team is involved in research and extension relevant to the Huang-Huai basins. Two of the pilot projects to which teams B belong are sponsored by components of CAAS. However, these are demonstration projects that involved little scientific research as such. Approximately 100 people in each county are employed at the county or commune levels in research and extension work. The main functions of "research" at this level (rightly) are verification and adaptation of externally introduced seeds and techniques and their promulgation, i.e., extension work. At the county level, there are usually an Agricultural Research Institute or Agroextension Station and stations for seed improvement, plant protection, agricultural machinery, animal husbandry, and forestry, each with five to fifteen employees. Typically each commune also has an Agrotechnical Station with one agronomist in charge. Experimental groups exist in some brigades and teams, which are involved in demonstration trials and seed multiplication. This "four-level research network" (county, commune, brigade, and team) utilizes meetings, short training courses, and publications as its main vehicles for extension. Usually personnel include a handful of college graduates, heavily concentrated at the commune level, and a larger number of middle-school graduates. The research organs in this region lack laboratories and all but the most elementary equipment; they are not equipped to fulfill the important functions of monitoring and testing. The quality and quantity of their training also leaves much to be desired. Each commune maintains 5–10 hectares and each brigade 1–2 hectares of demonstration and/or seed multiplication fields.

On the whole, this structure serves well to mobilize farmers for action and to provide some basic services. The weaknesses lie with the knowledge that is extended: very little meaningful or novel research has been done since the 1950s at any level, and understanding of scientific research techniques is concentrated at the top of this network. Forms of research that are particularly lacking are those that would make it possible to distinguish which of any set of alternative techniques is more or less effective and economic.

*Seed*

Provision of improved seeds in each county is now the responsibility of the county seed company. Breeder seeds from the provincial Academy of Agricultural Sciences or the county Agricultural Research Institute are multiplied either by county-managed seed farms (with very limited acreage) or else by commune production teams or scientific experiment teams under contract to the county seed company. The seeds sold are reasonably well received, but they represent under 10 percent of annual production requirements. Most seed requirements are met by farm selection and retention or exchanges among production teams. As a result, standing crops in this area show tremendous mixing of varieties, which entails crop losses as high as 30 percent. If company seeds were free of varietal mixing, their superiority would be easy to demonstrate and demand might be higher. This would require closer supervision of the seed production process, especially of subcontracting teams. As matters stand, the quality of commercial seed probably is not sufficient to compensate for its higher cost, and some seed companies have had trouble finding a market for expanded production.

## Fertilizer

Supply stations of the Supply and Marketing Cooperative are established at each commune and are responsible for the provision of chemical fertilizer and other inputs to production teams. The county allocates as much fertilizer as it has available (supplied by the province or county factories); unlike some other parts of China, in this area there is no standard allocation tied to acreage in each crop. Because fertilizer is volatile, it must be purchased within four or five days by teams to which it has been allocated; otherwise other teams in the same brigade may purchase it. Conversely, a team short of fertilizer must knock on factory doors to scrape up additional amounts—there is no alternative marketing system. Some fertilizer does find its way onto the free market, and private individuals may purchase surpluses from the Supply and Marketing Coop stock.

The soil in these basins is generally deficient in phosphate, and supplies do not meet requirements (this has become a nationwide problem that has not been resolved by the central government). Until recently, there were even brigade-level phosphate factories, depending on raw material imported at high cost from Yunnan and Szechuan. Teams fortunate enough to belong to a brigade with such a factory received ample phosphate, while others obtained little. Most of these tiny factories have now closed for want of raw materials; nevertheless, county phosphate factories keep on producing and providing the bulk of phosphate supplies, despite the inefficiency of transporting raw material instead of the finished product from the southwest.

## Agrochemicals

Pesticides are readily available in this region through the Supply and Marketing Coops and are used in quantities as high as 135 kg/ha for crops, like cotton, that are very susceptible to pests, but less than 10 kg/ha for most foodgrains. Unfortunately, DDT, "666" (BHC), and similar broad-spectrum and environmentally harmful insecticides make up the bulk of pesticides now employed. Effective but less harmful substitutes, such as Furadan, must be imported from abroad and are in short supply.

*Irrigation*

The "effective" irrigated area is only about 30 percent in the counties studied; the area that can be irrigated in years of drought is lower still. The supply of surface irrigation water is limited and cannot be expanded; in drought years, water supplied from North China's rivers severely contracts. Many areas do have underground water, but the amounts that can be safely exploited are uncertain—increased exploitation and severe drought in recent years has seen a drop in the underground water level in some areas. Coordination and control of irrigation facilities seem to be lacking: many more wells have been dug than equipped, and the area served per well is often less than half the design standard. Well-founded estimates of crop water requirements are missing, and little control is exercised over ground-water use. Organizational problems apply equally to drainage facilities: frequently main drainage channels are unaccompanied by the smaller channels that would make them useful, or the subsidiary channels have been deliberately plugged up to retain rather than drain water, rendering them ineffective. Construction and maintenance of irrigation and drainage facilities has become a more expensive proposition now that government policy changes and the disintegration of the production team have made it difficult to obtain corvee labor.

*Credit*

Almost all teams, and now households, must obtain credit to cover costs of input purchases. A portion is extended automatically, as with irrigation and electricity fees which have been deducted from team receipts (in a combination of grain and cash) after the harvest. The remainder must be borrowed from the commune's Credit Cooperative. The amounts loaned for working capital will generally not exceed 1,000 yuan, at an annual interest of 4.3 percent for a one-year term. As the Credit Coop has been the repository of all team receipts and working capital, as well as individual deposits, it has had virtually complete control over rural finance. The only alternative source of credit was an interest-free levy on team members, which teams sometimes used for equipment purchases. Teams with strong ability to repay—i.e., wealthy teams—could also borrow for longer terms (three to five years) at lower rates of interest (2.2 percent) for the purchase of farm machinery. Usually the Credit Cooperatives limited their lending to their own funds, another factor which discriminated against teams in less well-off communes.

The Credit Cooperatives are supervised by the Agricultural Bank of China, and in theory they may obtain funds from the latter for relending. The bank was only reestablished in 1979 and does not operate below the county level in this area, although it plans to open offices at the commune level as part of its planned expansion of activity in the 1980s. In general, the shortage of rural credit is probably the most severe constraint on development. Moreover, as my use of the past tense in describing the role of the Credit Cooperative implies, a wrenching readjustment may be required for this institution to serve adequately the new household-based farm economy. Since each household will prefer to own its own draft animals, tools, and equipment, the demand for credit may explode.

## Farm Machinery and Repair

The rate of growth of machinery use in the basins has been high in the past, with emphasis on pumping equipment, tractors, stationary threshers, and processing equipment, in approximately that order. Tractor-plowed area allegedly ranges from 15 to 50 percent, but actually may be much lower. Mechanization was previously promoted by subsidizing inefficient local farm-machinery factories producing low-quality equipment, which now sit largely idle for want of market and raw materials. Soft loans or grants were extended to encourage machinery purchase, with assessment of needs based only on the reported percentage of specific mechanized operations.

The demand for machinery in the area is limited by poverty and lack of credit, now that the system of virtually giving away machinery has been ended. The potential for mechanization would appear greatest during the June crop-turnover period, when timing is sometimes critical. However, the common practice of sowing maize between the rows of wheat before harvest, which makes it possible to squeeze two crops into a short growing season, precludes mechanized wheat harvesting or maize sowing. Moreover, teams prefer to minimize cash costs when labor-intensive methods will suffice, so they often leave machinery idle to save operating costs. The new system of farm management has accentuated this concern and changed the whole nature of the market for farm machinery, as discussed above.

Machinery maintenance and repair is also a major problem. The county Agricultural Machinery Company is the only source of spare parts, which are frequently not available on short notice. The communes maintain tractor repair stations, and repair of other machinery can often be done at the commune or brigade levels, or by inviting their repairmen to the production team. Minor repairs are often done by the teams themselves. But either poor-quality manufacture or poor maintenance leads to frequent breakdown. For example, the bearings of the diesel engines that power the popular 12 hp walking tractors must be replaced about once a year. Ningling County Team A was unable to irrigate from its one powered well during the 1981 drought because the pump broke twice in succession, followed by a breakdown of the engine that powered it; each time the inability immediately to obtain parts delayed repairs.

## Conclusions

As the comparison of the sample of production teams indicated, there is considerable development potential in North China agriculture that can be realized through government assistance. In the past such assistance as the Chinese government was willing to give was often squandered because of inadequate financial incentives, deficient planning, organization, and management, and weakness in the delivery of farm services. Recognizing these lacunae, the government since 1978 has moved to increase farm prices and adjust their structure. Because collective farming has had little to offer in these impoverished areas, the government has permitted—even encouraged—a return to a household-based farm economy. It has also at least begun to buckle down to the more tedious task of upgrading the farm services required to support agricultural development. Thus the prospect in the rural areas of the North China plain has probably never looked so promising.

The new structure of farm prices seems to provide the proper incentives for choice of crops and use of productive inputs by all but the poorest farms. However, many farmers may still consider the price level too low to make farming a profitable occupation. This is not an unusual complaint, either in the history of Chinese agriculture or that of other countries. Further increase in the farm price level, which the government would prefer to rule out, is not the only nor the best way of improving the situation. An augmented supply of better and cheaper consumer goods and expanded employment opportunities in rural industry and commerce would indirectly increase rural farm incomes and improve rural morale at less cost to the nation as a whole.

Abandonment of collective farming in these areas is surely the most radical measure the government could have taken to improve farm management, and this alone will help to increase the efficiency of labor and capital use, thereby raising incomes. By itself, it is doubtful that this reform will have much effect on crop yields—my supposition is that collective farming got the job done, if in a wasteful way. But the reduction of waste does free up labor and capital for other uses, and some of these uses will be agriculturally productive.

It is hoped that this reform will come to be accepted as permanent. Otherwise local government will not accomplish the restructuring of agricultural services to serve a household-based farm economy without which fruits of reform may be meager. It seems likely that provision of these services will continue to be the responsibility of government organs, so that the reform of public-sector management must be tackled. Beyond organizational reform, many of the current weaknesses of agricultural services call for investment of government funds. In some cases, such as with agricultural credit, this is direct investment in the agricultural sector. In others this is investment in the institutions or industries providing support services or products.

With the concentrated application of political power, the issues of price reform and restructuring of farm management could be accomplished with the aid of a central directive or two. Not so the reform of rural public-sector management and the investment program required to upgrade farm support services. The fragmented and long-term effort required will pose an even greater challenge to government leadership.

## Notes

1. Lin Shen, "The Inside Information on China's Economic Readjustment," *Zhengming* (May 1979):11.

2. For example, Ningling A has only one pump for six wells, and yet its irrigable-acreage figure requires six wells operating at maximum efficiency.

3. As costs of collective pig raising exceed revenues, this is of benefit to the state rather than Team B; it is a "hidden tax" on the more prosperous team.

4. Thus the structure is assessed from the prices of crops taken relative to each other and/or to the prices of farm inputs; the level, from the prices of farm products relative to industrial products (other than farm inputs). Each set of relative prices can vary with no change in the other. Also, in assessing the structure, *marginal* prices are most relevant, whereas *average* prices are compared in considering the price level.

5. It is the difference in intensity that leads to the apparent disagreement between my conclusions and those of Steven Butler in this volume. His village is precisely one of those model, intensive-cultivation villages where grain production was probably unprofitable at the time of his study, much as he argues. However, the problem was not in price but in intensity.

6. In fact, above-quota "prices" don't exist as such—rather, percentage

premiums are applied to the totals of sales exceeding the quotas for certain categories, e.g., grains. Otherwise there would be an incentive for teams to use their low-quality, low-priced grain to satisfy quota requirements and reserve their higher-priced grain for above-quota sales.

7. Prices of imports or exports adjusted for transportation cost and processing to the same geographic location and standard as farmgate prices.

8. In "Agriculture in the Four Modernizations," *The China Geographer*, no. 11, *Agriculture* (Boulder: Westview Press, 1981).

9. Locally, at least, "*baochan daohu*" (HRS) is equated with "*da baogan*," which virtually eliminates the role of the team.

# 4

# Price Scissors and Commune Administration in Post-Mao China

## Steven B. Butler

After 1978, China's rural organization moved rapidly through a series of radical structural changes—starting with higher prices, increased openness to commercial crops, and ostensibly less administrative intervention; continuing on through systems that increasingly approximated family farming; and today involving an attempt to weaken the commune and its party organization as an administrative unit and substituting in its place economic management organizations. These developments can be seen as support for the view that China's collective agriculture was a great failure.

There is some truth to this view. But there is much more that must be said if one is to understand how peasants responded to collective agriculture in the past, the reasons reform is difficult to stop at a halfway point with part-market/part-bureaucratic control mechanisms, the reasons there will be temptations to return to bureaucratic control mechanisms in the future, or why much direct administrative control continues to exist in many ways even today.

Some of the pressures faced by Chinese agriculture are not specific to China or to socialist agriculture but are faced by any society with a state bent on rapid industrial development. In these societies urban interests are likely to dominate and peasant interests to suffer. Administrators faced with implementing state policies of less benefit to peasants are placed in a difficult position.

Some of these themes emerged in the previous two chapters, but this chapter illustrates them with richer detail on how a single locale coped with both the trends of the previous thirty years and the initial reforms of 1978–79. The evidence presented here shows that for many years China's peasantry endured, even accepted, unfavorable terms of trade with the state, increasing grain production while earning less profit.

Western and Chinese observers now criticize the managerial inefficiencies of collective agriculture as well as the economic inefficiencies caused by the price structure. Yet peasants seem to have accepted many aspects of the collective system. Research on peasant societies suggests that the need to secure subsistence is of overriding concern to most peasant families over and above maximizing income or pursuing rapid development.[1] I argue that concern for guarantees of subsistence caused peasants to be more tolerant of a system that in many ways performed badly from an economic standpoint. The system did effectively ration available food. Dahe Commune's apparent history of strong leadership and generally good management may also have made the system more palatable than in other areas. But recent improvements in food supply as well as increases in cash income make the threat of starvation and destitution much more remote for many peasants in China, especially for the younger generation in relatively prosperous regions. Their willingness to accept an administered system may depend much more than in the past on improvements in efficiency and on rapid growth.

This chapter is based principally on six months of field research during the spring of 1980 in Dahe Commune, Hebei province, a time when many administrative reforms were being introduced. Extensive discussions were held with commune and brigade leaders about the underlying managerial problems of Chinese agriculture and the intent of the reforms. Data from this half year of field research are supplemented by data drawn from interviews in Hong Kong in 1977 and also by data from the Chinese press.

## The Economic Background

In 1962 China abandoned in name a "Soviet" model of economic development that stressed rapid growth of heavy industry financed by large extractions from the countryside. The government announced a new ordering of developmental priorities—agriculture first, followed by light (consumer) industry and finally heavy industry. Agriculture was to be modernized with increasing use of machines, chemical fertilizer, and high-yield seed strains. The agricultural tax—a fixed amount in kind—remained frozen, allowing agricultural collectives to retain a larger percentage of their produce for sale to the state or for internal consumption. Agricultural prices, too, had increased several times, improving the price ratio of agricultural and industrial goods in favor of the agricultural sector. By all appearances the government was making a serious effort to beef up agricultural production and satisfy the economic demands of China's huge peasantry. If collective agriculture was a workable system, surely China's new policies would give it a chance to prove itself.

Unfortunately, the reality of the economics did not live up to all the lofty policy goals. A brief tracing of economic trends in Dahe Commune since 1958 should illustrate fairly well the crunch that many communes experienced. I believe that Dahe can be taken as broadly representative of moderately prosperous grain-producing regions of North China.

Figure 4.1 shows trends in the yields of multiple-cropped grain in Dahe, principally wheat and corn. Actual yields, for a variety of reasons, are somewhat smaller than the figures used here, but the figures reliably reflect trends. The year 1958 was a good one in Dahe. Yields reached a historic high of 2.91 tons per hectare, and per capita production hit 252 kilograms. The combined effect of the administrative disaster of the Great Leap Forward—which local cadres assured me was the principal culprit—and bad weather took its toll in subsequent years. Grain acreage was cut while yields dropped steadily, nearly 25 percent by 1961, to 2.42 tons per hectare. Per capita output plunged to 296 catties, 148 kilos per person. After paying taxes and completing compulsory sales to the state, the commune did not have enough calories to go around. The birth rate declined to nearly zero. Miscarriage was common and infant mortality high. Registered population in the commune declined despite the return of many former villagers from the city. Yields increased from 1964 to 1965 and then stagnated for the rest of the decade.

Beginning in 1970, the commune began to increase dramatically its grain yields. From 1969 to 1975, yields more than doubled. During the decade grain production in Dahe experienced "modernization." The commune dug and improved canals connecting it to the Gangnan Reservoir system. Production teams sunk many new mechanized wells. Applications of chemical fertilizer increased annually, and the commune introduced new seed varieties and new cropping systems. The reigning agricultural slogan—"take grain as the key link"—received careful attention in Dahe.

But the rapid increase in grain yields did not have such a positive effect on distributed income. Personal income, as reflected in figure 4.2, fluctuated erratically during the 1960s, stabilized in the early 1970s, and finally in 1977 began to increase rapidly. (The figures used

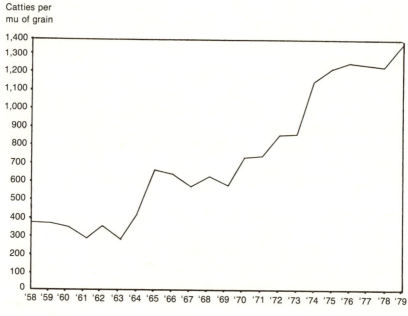

Catties per
mu of grain

Figure 4.1

Annual Grain Yields, 1958-1979

here include the cash value of goods distributed in kind.) The decline in income in the early 1960s clearly reflected the disasters of the Great Leap Forward. For the rest of the decade, however, net production team income (as opposed to distributed income) was not nearly as erratic as it might appear. Dips and turns on the line for distributed income for these years tended to move inversely with the rate of savings (figure 4.3), which in turn followed the capricious policy changes imposed by higher levels of government. Family income in these years did not bear a close relationship to production team income.

What is surprising is that in the six years between 1969 and 1975, when yields doubled, distributed income, as well as net production team income, stagnated. But two years after grain yields basically leveled off in 1975, distributed income began to rise dramatically. This odd relationship is explained in figure 4.4, which expresses a ratio between team gross production expenses (which does not include labor costs) and gross income. The figure rises from .27 in 1969 to an incredible .52 in 1976, and it falls to about .46 after that. "Modern" farming, especially chemical fertilizer, was extremely expensive. In fact, pro-

Yuan

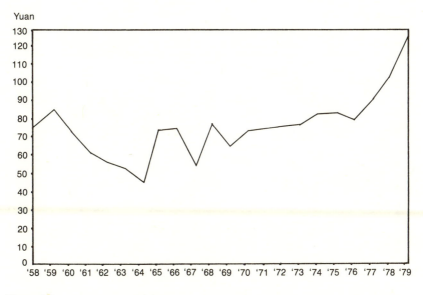

Figure 4.2

Annual Per Capita Distributed Income, 1958-1979

duction teams earned less money with multiple cropped yields near 10 tons per hectare than they did when yields were a more modest 5 tons per hectare. Commune members did benefit from the higher yields since they received a larger food distribution in kind, and more of it in wheat, which they vastly prefer over corn. But as food distribution went up, families actually received less cash each year from the collectives, forcing them to rely more heavily on family sidelines to make up the difference. The increases in cash distribution since 1977, which commune members are very happy about, come not from agricultural income, but from income earned in collective sideline industries, such as brick kilns, casting shops, noodle factories, and sewing groups, and, in some teams, by a reduction in the rate of savings. Including commune and brigade-level enterprises in 1979, about 50 percent of the commune's gross income derived from the more profitable nonagricultural sources (which in the local accounting system included fruit orchards and fisheries).

Production teams obviously did not work so hard to increase their grain yields with the idea that they would earn a fat profit. Initially, peasants may have assumed that modern agricultural inputs would pay for them-

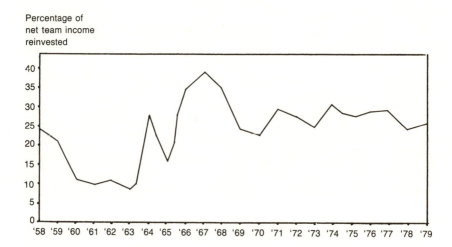

Percentage of
net team income
reinvested

Figure 4.3

Rate of Production Team Collective Savings, 1958-1979

selves, but after several years it became clear to all what was happening. Why, then, did teams continue to work so hard?

Clearly one incentive in Dahe was the improvement in diet. As wheat yields in particular have risen, peasants have had to eat less of the detested corn meal and have filled themselves on wheat noodles, steamed bread, and dumplings. Dahe sells only 25 percent of the grain it produces to the state (the national average is under 20 percent). The economy is thus still largely subsistence, and even though the amount of grain sold to the state has steadily gone up, as a percentage of gross yields it may have dropped slightly. Excess corn goes to feed pigs. Collectives tend to lose money on pig raising, but families may stand to gain if the ratio of pigs to family members does not grow too high, since table scraps and other waste make up part of the pig's diet, while the collectives cultivate food specially for the pigs. The lower cash distribution received by team members is at least partially compensated for when they sell a mature pig to the state. And meat consumption is up as well.

As should be apparent, all the economic calculations, all the pluses and minuses, become quite complex. They are so complex that peas-

Ratio of
production expenses
to net income

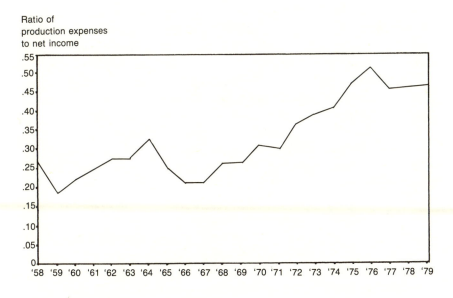

Figure 4.4

Cost of Production Ratio, 1958-1979

ants, production team heads, commune cadres, and higher level officials do not try to figure things to the last penny. In fact, using 1979
methods of production team accounting, it was virtually impossible to
calculate production costs in any meaningful way, even if team accountants tried, which they did not. Accountants did not record fertilizer,
insecticide, water, or labor applications for individual crops, which
would have been doubly difficult due to interplanting. Losses in pig
breeding or draft animal management would be partially offset if the
agricultural accounts recorded an expense for manure or plowing, but
in the absence of an open market no one was sure how much to record,
and it would have been a lot of bother anyway. When I asked team
accountants why they did not bother to figure production costs, I
generally received a response that went something like this: "What
difference does it make? The state tells us how much to plant of each
crop. They advise us on planting methods, and they tell how much to
sell to them. We can't make much money in agriculture anyway, and we
have to eat."

In spite of the fact that accountants did not try to calculate carefully the exact cost of producing grain, they did tell me that they lost money on grain sales to the state. In 1979, when higher authorities gave teams the right to decide how much chemical fertilizer they would apply, teams generally applied less, even though the price was cut. When the conversation reached this point with several accountants and team heads, I frequently popped the question whether it would not be better if the state raised grain prices further, and would not grain cultivation become even more burdensome in the future if prices were not raised? The answer was surprising. Many just shrugged their shoulders and said that while of course it would be nice to get a better price, the state had just raised prices and it could not afford to do so again. In any case, they felt, the teams had other ways to increase their income, most of the grain was for their own consumption, and their diet was steadily improving. The findings on this question during my stay in Dahe are largely consistent with reports from refugee informants in Hong Kong describing communes in the wealthy Pearl River Delta. They reported that the most contentious issue there was not the basic obligation of teams to plant 80 percent or more of their land in rice, but restrictions placed on development of sideline enterprises in some locations and restrictions on the size of the grain ration. The state enforced these unpopular decisions through heavy, top-down administrative pressure.

In other words, production teams seemed to have accepted or are at least resigned to a general responsibility to sell grain to the state, earning little or no profit as long as team members' diet continued to improve and income went up. James Scott's observation that in peasant societies what is left is more important than what is taken away seemed to be true.[2]

Nonetheless, by the late 1970s, the government had concluded that the agriculture sector was in crisis. For twenty years, per capita grain consumption in China stagnated despite superior performance in places like Dahe. The pressing need to produce grain had caused local administrators to discourage villages from developing sideline industries. Although the policy was applied unevenly, the effect in many areas was to choke off avenues for teams to increase distributable income. The failure to rise generally, along with the use of compensation systems that did not always reward hard work, caused many peasants to become demoralized.

The response to this crisis prior to 1979 had been to clamp tighter administrative controls over agricultural production. The controls in many cases looked very clumsy. Administrators occasionally forced teams to use agricultural techniques that did not work. In Dahe, county administrators forced teams to interplant wheat with sorghum for several years, even after much evidence had been amassed that maize would produce higher yields (this was not true in hilly regions nearby). In 1976, during a winter drought, county officials ordered teams to flood their fields with well water, a technique that had been tried with success farther south. But in Dahe the water froze promptly and killed much of the winter wheat.

Judging from the reports of Guangdong peasants and the inability of Dahe officials to come up with more examples, these gross errors seem to be the exception rather than the rule. But over the years teams grew into the habit of looking up the administrative ladder for instructions on what to do. Many team leaders did not take a great deal of initiative in solving their team's economic and technical problems.

Peasants and team leaders now complain that their methods of production have changed so rapidly and become so complex that they do not fully understand the interrelations between all the multifaceted aspects of production. As the vice head of the Institute of Agricultural Economics, Mr. Wang Gengjin, said to me, "Peasants know how to farm with traditional methods. But they do not understand how to farm with modern methods." This situation seems to result from the low level of education, inadequate price signals, and the fact that technical change has been imposed from above rather than learned and voluntarily adopted from below.

The habits of looking above for instruction, the headaches of running a team well, and skepticism about whether better methods of agricultural production would result in improved life in the villages combined to produce a malaise in many communes. Peasants who wanted to improve their lives turned their attention elsewhere. Undoubtedly there were many exceptions to this general condition, a few bright spots here and there where a particularly able leadership core made a big difference. But for each of these spots of prosperity, there were more villages of stagnation or, worse, desperate, unspeakable poverty.

## Reforming Labor Management: The Responsibility System

After 1978, China's leaders tried to encourage a change of attitude. The change in the broadest sense fell under the slogan "seek truth from facts." The slogan represented a new more pragmatic attitude and intent. Policies and programs were supposed to live or die according to whether they worked and not according to their ideological premises. In the economy, accordingly, "economic management" methods were to substitute for "administrative management" methods. At the ground level, the new attitude boiled down to changes in several management areas. Broadly, the changes were designed to strengthen material incentives for peasants and local cadres, to enhance unit decision-making autonomy, and to "rationalize" management in general by fully specifying important management goals rather than surrendering to the rule of vague political slogans.

Production team heads in Dahe told me that prior to the introduction of a new "responsibility system" in 1979, it was difficult to manage labor. Dahe used the "Dazhai" system of workpoint allocation right up until 1979. Workpoints recorded the contribution of labor of each member to the production team. At the year end they were totaled and used to determine each member's share of the team distribution, in cash and kind. Under the "Dazhai" system, laborers were graded and received a fixed allotment of workpoints for each day they worked. When the system was introduced in 1965, assessments for grading purposes were made frequently and retrospectively. But over time, the assessment process proved too time consuming and rancorous because team members frequently disagreed about the value of their contributions. Grading bore less and less relationship to any careful examination of actual work performed and the system lost its ability to discriminate among slackers and hard workers. Women, who gradually came to perform about 80 percent of all field labor in the commune, more or less automatically received a lower rating. Team heads complained that it was difficult to get people to show up on time for work, to perform quality work, or to stay until quitting time. The job of team head became unpleasant since, in order to do a good job, the head had to spend most of his time brow beating neighbors, friends, and relatives. When team heads tried, occasionally, to use other systems, they were rebuffed by higher authorities.

The new policies gave greater freedom to team leaders to reward team members more precisely. Team heads in Dahe began to use a combination of time, task, and responsibility systems depending on technical aspects of the work performed. A time rate simply awarded points for the amount of time spent at a specified task. The work performed in, for example, leveling a field was difficult to quantify in any other way. Task rates awarded points for the amount accomplished: the area plowed, the number of carts of earth moved, or the area of land irrigated. Task rates encouraged speed but offered no incentive to insure quality. The responsibility system awarded points according to the output, or results of labor, such as weight of wheat, corn, or cotton on an assigned plot of land, or the profit earned for the team while driving a tractor or mule cart for outside contractors. Although at the time of research local cadres had not adopted this terminology, the system appeared to be a combination of contracts for specialized side-line tasks (*zhuanye chengbao*), piece rates (*lianchan jichou*), and individual labor contracts (*lianchan daolao*), much as described by Parish and Zweig. The system encouraged efficiency and diligence but was complicated to administer.

The system adopted in Dahe Commune varied from village to village in its details. Although production teams were supposed to have autonomous rights to draw up their own system, teams in the same brigade tended to adopt very similar systems, indicating that brigade leaders continued to play an important role.

Corn production in Jiacun Brigade provides a good example of the complexities of the new system. After planting, each team member pledged to take responsibility for a piece of land. The brigade assigned an output target for each parcel. Agreeing on the target was the most contentious step because it was used to determine bonuses and penalties. Members received two workpoints for each catty of corn. For each catty over the target, they received a bonus of .05 yuan cash. For each catty under, they were docked one workpoint. The output quota was determined after examining the quality of the land, the development of the sprouts, and the previous year's output.

But the production teams did not just let team members manage their plots as they saw fit. A bell rang in the morning and members were expected to be out working. The team decided when to weed, how to apply fertilizer, and numerous other technical specifications as well. If a member did not perform quality work, the team head might assign someone else to redo the job and pay him or her with workpoints that would have gone to the one who normally managed the plot. If a member worked behind schedule, he might be penalized.

The system did not simplify record keeping. As a precaution, all work performed was meticulously recorded as though it were being used to calculate workpoint totals on a task-rate basis. Then if a natural or human disaster destroyed the crop, workpoints might nonetheless be determined.

In the case of wheat, however, technical problems made it difficult to use the responsibility system because of the difficulty of determining exact yields from small parcels of land. To obtain an accurate measurement, wheat must be weighed dry, because wheat from different fields may lose different amounts of water in the drying process. With the shortage of labor, threshers, and drying grounds during the critical few days when wheat must be cut, threshed, and dried, teams could not afford to keep the harvest from individual plots separate. Dahe Brigade used a system of estimating the yield of a plot of land immediately before the harvest. The production team plowed, planted, irrigated, and sprayed wheat fields using a task-rate system. Everything else from planting to harvest was rewarded by the responsibility system. But in Jiacun Brigade, it was decided that estimating yields would provoke too many arguments since it could not be verified. There, wheat production followed a task-rate system entirely.

The details of the system varied from team to team, and for different crops. With a new emphasis on increasing cotton production, some brigades adopted a sliding bonus scale for calculating points, building in sharp penalties and rich rewards depending on the yield achieved. A peasant in Xiaohe Brigade would earn only 3 points for each kilogram of unginned cotton at yields under 1.35 tons/ha. The rate climbed steeply to 2.5 points for each catty if yields surpassed 2.1 tons/ha.

Suanghe Brigade assigned two drivers to each of its tractors and asked that they give the team 10 yuan per day, 365 days a year. When they drove for the team, they received 18 yuan a day. Otherwise they hauled goods for outside units. The drivers paid for gas and repairs themselves, and pocketed any excess over the 10 yuan. The drivers also received 10 work points each day. In 1980 the drivers expected to earn about 500 yuan cash each, a very substantial amount. Various and complex systems were developed for draft-animal breeders and drivers, pig raisers, noodle makers, bee keepers, and for any other type of work where results could be measured.

Production team and production brigade leaders I spoke with all reported that the responsibility system had a dramatic effect on work performance. Since work quality affected income, quality improved. Some tasks, I was told, were now performed in less than half the time previously needed. If members finished work early, they might be assigned other jobs on a time or task rate. As the efficiency of agricultural labor improved, team heads assigned members to work on more lucrative sideline production. And team leaders reported that they could devote their time to managing and coordinating team affairs rather than hassling team members to work hard.

## Reforming Management: The Contract System

Accompanying the responsibility system was a new contract system. At the production team level, all the details of the responsibility system were written down and signed by team members and management. But more importantly, contracts began to be signed between different levels in the commune and between the commune and the county. In Jiacun Brigade, the brigade signed with all production team management committees contracts that contained targets for grain, cotton, edible oil, and pig production, as well as cash income from sidelines. If the team fulfilled all targets in grain, cotton, oil, and sidelines, the team committee received from the brigade a cash bonus of 40 yuan per committee member, which they divided among themselves. A 10 percent overfilling of the grain target earned an additional 10 yuan per cadre. For each 7.5 kilos per hectare over 472.5 of ginned cotton, the committee received 1 yuan per member. When the pig production quota was filled, it earned 5 yuan per member. If any single target were missed, the total bonus was cut 10 percent. If the targets in grain, cotton, or edible oil were missed, committee members received workpoints equal to the average of full-time laborers in the team. But if all the targets were met, they received a workpoint total equal to the average of the highest ten earners.

The targets and bonuses varied from village to village. Some villages had a target for a low birth rate. But regardless of the details, team cadres came to have a strong financial incentive to meet a broad range of targets. Brigade cadres, on their part, receive similar bonuses from the commune when they meet their targets. And commune cadres, who are on the state payroll, receive a cash bonus from the county. The system offers for the first time financial rewards for individuals for meeting targets in the plan.

Contracts are signed between units with obligations of each side clearly spelled out. In addition to paying the bonus, higher administrative levels are also enjoined to supply fertilizer, pesticides, diesel fuel, and other needed inputs in specified minimum amounts. I asked if the contracts had any legal status, whether parties to the contracts had means to redress disputes arising from obligations in the contracts. The question brought smiles. The "contract" is really more of a plan of action than a legally binding document. But cadres did not feel that compliance was a problem. Instead, they pointed to the benefits of the new system. Now, all important facets of production team work were clearly specified, not just the grain targets. Financial rewards clearly reflected the relative importance of various targets, and for the first time, the government clearly committed itself to provide needed inputs. For the first time, apparently, everyone was in clear public agreement about what they were supposed to do.

The contract and responsibility systems went a long way toward solving two severe problems: providing direct material incentives to motivate quality work, and clarifying exactly what is expected of each unit and individual. Previously, vague slogans too often left cadres uncertain of exactly what was expected of them.

### "Administrative" or "Economic" Management?

Oddly enough, the new responsibility and contract systems in some ways went against the thrust of other trends favoring decentralization, increased autonomy for production teams, and the reactivization of competition and controlled market forces. True, stronger material incentives were consistent with the general meaning of "economic management," but at this intermediate stage production teams came to have *more* rather than fewer targets to meet. By any common-sense understanding, these targets were the result of administrative decisions, not market forces. Was this consistent with moves to reduce heavy-handed bureaucratic controls over agricultural production?

Also, commune and county cadres were not finished telling teams how to plant their crops, and they may continue to tell individual families how to do so today. While at Dahe, I sat in on a long series of meetings with commune and brigade officials to review and discuss the maturation of the wheat crop and to plan for the sowing of corn. At the end of the day—most of it spent cycling around the commune to inspect the fields—we met in a closed room where the commune secretary laid out six "standards" (*biao zhun*) for sowing the corn, including how far apart to lay the seeds, how deep to plant, and how to apply fertilizer. The meeting ended with a pep talk replete with familiar slogans, and a reminder that all production teams must plant according to the standards.

After the meetings, I discussed with commune Secretary Shang Cunlu and Vice Secretary Zhang Yumin, who was responsible for agricultural production, the moves to loosen control over team management. I tried to push the two men, both of whom struck me as extremely capable, to explain how the new policies from Beijing affected their work in managing Dahe's economy.

"Last year," explained Shang, meaning 1979, "the State Council sent down two documents on agriculture. The basic structure of the commune was reaffirmed—three levels of ownership with the production team as the base. The changes brought about a reduction in the use of administrative orders in running the commune. There are fewer meetings, fewer report forms, and fewer reports. We now conduct more investigations at the basic level than before.''

Shang further explained that the production teams now had formal autonomy for drawing cultivation plans (although they had to meet quotas set by the state), formulating a distribution plan, deciding on technical measures to increase output, and formulating internal management rules. In addition, Shang told me, teams had the right to refuse to obey ''blind commands'' issued by higher level officials.

The six standards that Shang had earlier issued for the planting of corn seemed to me to be inconsistent with the reforms he had just outlined, so I pushed him to explain. I asked him if the six standards were not administrative orders, which he was to avoid. Shang explained hastily that an ''administrative order'' (*xingzheng mingling*) was the same thing as a ''blind command'' (*xia zhihui*) and was to be avoided. However, an ''administrative order'' should not be confused with an ''accurate command'' (*zhunchue zhihui*). Shang's six standards were an ''accurate command'' rather than an ''administrative order.'' Vice Secretary Zhang did not agree with this hair splitting of meanings, and the exchange that followed is worth quoting at length for what it reveals about the underlying managerial problems in Chinese agriculture.

''We cannot avoid the use of administrative orders,'' explained Zhang. ''There are a lot of production teams leaders. They change frequently. Many of them are good and understand how to run their teams well, but some are not and we must issue orders to them. We must give them annual planting plans. Otherwise they would not fulfill the plan and would plant other crops. Team heads vary in their abilities. If they don't lead the teams well, we must take charge and lead them.''

''If we set standards like we did yesterday,'' said Shang, ''and the leaders do not accept them, we must increase our educational efforts. If they refuse to implement them, they must have good reasons. They must have concrete justification for what they do. We would not let the teams plant corn seedlings eight inches apart instead of the standard of from five to six inches.''

''We would not let the team plant at eight inches,'' agreed Zhang, ''and would make the team redo the work. The team must have objective proof if they go against a standard like that.''

''The difference between an accurate and an inaccurate order is not always clear," said Shang, "since conditions change from year to year. Therefore, we must be in constant touch with reality."

"We first use reason," said Zhang, "and try to persuade the teams and brigades what to do. But if that doesn't work, we must issue an order. Not all the team leaders have the same level of understanding and we cannot let their mistakes affect the collective economy. That is too important. It affects the members' lives and their enthusiasm for the collective."

Shang concluded: "The production teams have autonomy now, but this does not mean that we can be lax in our leadership. These two aspects must not be seen as conflicting. They are consistent and we will continue to lead the teams vigorously. But our opinions and leadership must be based on reality, on the masses' experience."

Despite the technical disagreement between these two men about what constitutes an administrative order, their views about how to run the commune were consistent. They believed that the production brigades and teams could not simply be left to manage their own affairs. Teams needed to be watched, checked, and prodded to do what they were supposed to do. The general understanding of what teams were supposed to do was much broader. Economic planners tended to view teams as grain producers for the state. Now, after long debates in the official press, a consensus was reached that the purpose of production was actually to improve people's lives. As such, increasing cash income was said to be as important as maintaining and increasing grain yields. But this decision was not left to the teams to decide for themselves. Teams probably did have more real authority to plan their own affairs, but this increased autonomy occurred in a context in which they were still heavily subject to regulation from above. Bureaucrats in China found it impossible to relinquish the use of administrative measures to run the economy, and they hesitated on using entirely economic measures.

## The Impact of the Price Scissors

Was the hesitancy to move further toward "economic management" the result of bureaucratic foot dragging and conservatism? Or were there sound reasons for going slow?

The dilemmas facing Secretary Shang and Vice Secretary Zhang highlight nicely some of the problems that continued to plague Chinese agriculture. Superficially, the problems appeared to be managerial in nature, and that is how the Chinese previously treated them. Shang and

Zhang were concerned about the very real problems of team heads who mismanaged their teams or accountants who could not keep their books straight. Production teams were complicated, multitask organizations that required skill to manage. Shang and Zhang might blame these problems on the Gang of Four or on the low quality of education in the commune. If left unattended, members of a poorly managed team would become discouraged. Even if collective agriculture was a good system, in the best interest of all members, bad management aroused mistrust among members and caused them to go their own way. It would be difficult to keep such an agricultural collective together, working hard and efficiently to improve yields. These were not imaginary fears for Shang and Zhang but problems that were only too real.

But why, we must ask, did management of the teams prove to be such a headache? Why couldn't good leaders be found and encouraged to stay on the job? Why were peasants uneager to learn accounting and management skills needed to lead the teams effectively? The reason, it would seem, was that most peasants continued to regard agriculture as a dead-end occupation. Thirty years of propaganda glorifying the virtues of field labor seemed not to have altered this view. In Dahe, nonagricultural work usually earns fat cash bonuses unavailable to those who labor in the fields. The bonuses reflect an overall price structure that makes nonagricultural enterprises much more profitable. Men have gradually moved into nonfarm jobs, and women in Dahe now perform 80 percent of all field labor. They can earn many work points, but when totaled up, their income will not equal that of a man who drives a tractor or quarries stone. Rightly or wrongly, young peasants see the reforms in education as giving them a green light to study hard, enter schools outside the commune, and leave the village for a secure state-paid job. Cadres told me that recent increases in income and other improvements in the quality of village life somewhat dampened this fever to get into the cities, but the fever still burns.

The root problem lies in a price structure that, despite reforms, continues to make agriculture economically unattractive. In 1979, the state boosted agricultural purchase prices by over 20 percent. Those teams prosperous enough to have grain to sell to the state enjoyed an immediate windfall. Peasants in Dahe were very happy about the extra income, and happy as well that finally the government seemed to recognize their demands for a better life as legitimate.

But even though the price hike improved the balance sheet, it did not alter fundamentally the fact that farming does not pay.[3] One team head told me he figured that before the price hike his team lost money on grain cultivation. Afterward, he said, the team broke even. In 1979 and 1980 the government tried to relieve the tremendous pressure placed on teams to expand grain acreage and increase yields. Newspaper articles criticized the slogan ''take grain as the key link,'' and they encouraged teams to diversify their crops and economic endeavors. As a result, in 1980 grain acreage declined and the state found itself short of grain. (The shortage was aggravated by bad weather as well.) In 1981 the press took a more curious stand, encouraging diversification, but most Chinese leaders regarded foreign purchases as only a short-run solution. They continued to feel that the international market was unreliable and did not want to waste precious foreign exchange. More voices suggested that China needed again to increase grain purchase prices to encourage teams to produce more. But the government was reluctant to raise the price of grain which it sells to urban workers. After the 1979 price hike, the state began taking a tremendous loss on grain trade. Given an already large deficit in the national budget, the government could ill afford to increase prices again.

What recourse, then, did the government have to insure adequate grain supplies? The most radical alternative, and the one finally chosen, was to separate the commune's governmental functions from its economic functions, so that only local cadres paid from the commune's own revenues would manage the collective economy. This move, it is argued, will promote greater responsibility on the part of commune leaders and prevent meddling in the collective economy by cadres on the state payroll whose personal fortunes are not linked directly to the commune's economic performance. There is the further argument that units of collective agriculture can function well only with the commune structure dismantled.

Already touted in 1980, these ideas were viewed by some people with whom I spoke in China as nothing but the fanciful spinnings of armchair theoreticians in the Academy of Social Science who had little knowledge of the practical problems involved in managing a commune and who believed that most commune leaders were incompetent. In fact, the root problem was not the ''meddling'' of men like Shang Cunlu and Zhang Yumin. Given the existing price structure, if the state did not have cadres around to enforce the plan, it would quickly discover that peasants would not produce what the state needs to buy. What is remarkable is not the clumsiness or bumbling of the system, but that it has performed as well as it has given the unprofitability of agriculture. Is there any other country in the world that could achieve, as it did in Dahe, a doubling of yields in six years, while rural producers actually earned less money?

## Conclusion

Much of the "malaise" in Dahe was effectively stemmed by reforms introduced in 1979 and 1980. The responsibility system strengthened material incentives, as did the new contract system, which clarified the organizational goals of each level of ownership in the commune. The new management methods were accompanied by a price boost, and family income increased. Villagers were encouraged to develop profitable sideline enterprises, which increased cash income even more.

But as much as the dramatic improvements in village life have satisfied immediate demands, they also aroused expectations. For probably the first time in history, modernization, as an idea, as an ideology, sank into the minds of Dahe's peasants. They came to believe that progress was possible and they expected life to improve each year. This, I was told, was quite different from popular attitudes in the 1950s, when peasants looked to the new agricultural collective not for progress or prosperity, but for security, for a guarantee against hunger and destitution.[4]

The short-run effect of the reforms was to inject new life, new vigor into the administration of China's collectives. But the question remained whether in the long run growth and enthusiasm could be maintained, and if not, what effect it would have on rural organization.

Production teams looked to nonagricultural sidelines to provide more income. But already, many villages were at a loss about where to expand. Small villages did not have the manpower, capital, or technical skills for many of the larger, more profitable small industries. Many brigades established brick kilns, but with cuts in capital construction on the agenda, local cutbacks in construction caused many teams to earn less money in sideline activities. There was not as much business for tractors and mule carts hauling brick and stone to construction sites. The common stone quarry sold far less stone and gravel. Commune leaders talked about ideas that they would spin their own silk, can their own fruit, and process their own cotton, but in 1980 so many communes began to process their own produce that many state factories ran short on supplies. Now the government wants to control the growth of such small factories.

But even if growth could be maintained by relying on sideline enterprises, it is difficult to believe that in the long run peasants will not just try to cut their losses in agriculture rather than concentrate on improving productivity. The process has already begun with first production teams and then families taking their best, most capable workers out of agriculture and putting them into more profitable sidelines where their skills can make a difference.

In Dahe, as in some other places, agricultural field work is increasingly the realm of women, the old, and the weak. These are trends suspiciously similar to those in the Soviet Union where the most capable and able-bodied parts of the labor force have abandoned ordinary agriculture.

Recent reforms, releasing more products from state control and providing higher prices for above-quota sales, increase the profitability of agriculture. But in poorer areas with few above-quota sales or in areas poorly suited to commercial crops, problems needing administrative intervention may still remain.

Indeed, even after the late-1984 price and planning reforms, ten agricultural products such as grain, cotton, vegetable oil, and tobacco remain within the state plan. And for these sorts of crops the individual farmer continues to sign detailed contracts with local leaders on the acreage and output of each crop. This leaves ample room for leaders such as Shang Cunlu and Zhang Yumin to continue to give direction in agriculture, and for farmers who are unaccustomed to operating in a market with more free-floating prices and uncertain future demand in some crops this direction may not always be unwelcome. Administrative direction has been reduced in intensity and scope, but it is far from dead.

## Notes

1. Despite their disagreements, this is a common theme in James C. Scott's *The Moral Economy of the Peasant* (New Haven: Yale University Press, 1976) and Samuel L. Popkin's *The Rational Peasant* (Berkeley: University of California Press, 1979).

2. See Scott, *The Moral Economy of the Peasant*.

3. Editor's note: Our authors are in less than full agreement on this issue. Wiens gives a more positive evaluation of current prices, suggesting that the problem in Butler's villages lay more in the intensity of cultivation pushed upon them from above.

4. To reiterate a point made briefly at the outset, there is no reason to believe that the price reforms alone have benefited measurably the poorest of China's villages, which may contain over 200 million peasants. These villages do not sell enough grain to the state to have gained from price hikes. Some of these villages may benefit from the new freedom to develop sidelines, but many have great difficulty marketing their products because of poor transportation facilities.

# PART TWO

*Changing Incentive Systems*

# 5

# Remuneration, Ideology, and Personal Interests in a Chinese Village, 1960–1980

## Jonathan Unger

Inefficient and poorly motivated labor has dogged socialist farming from Cuba to Algeria to Russia, in part because the socialist economies have found it difficult to devise rural payment systems that are at once economically viable and ideologically palatable.

In wrestling with this problem China, more than any other socialist country, experimented during the 1960s and 1970s with a wide variety of remuneration schemes. It will be seen here why each of these different wage systems was tried. Attention will center particularly upon what was called the Dazhai system, which comprised an explicit effort to reshape behavior in China's villages toward ends that Mao and his followers deemed morally superior. In the short term, the Dazhai system succeeded in some villages. But the program, by forcing peasants publicly to judge one another's attitudes and work performances, was eventually undermined by growing tensions among the peasantry.

To examine the shifts in attitudes among the peasants, I shall draw upon the experiences of one Chinese village. Located in the southern province of Guangdong, Chen Village contains approximately 250 families, about average in size for its district. It is not particularly well off, but neither is it noticeably poor by the standards of Guangdong. It has never been promoted by the government as a political or economic showcase for other villages to emulate, but neither was it ever considered politically backward. It is simply a village that, by happenstance, became the subject of study by two other sociologists and myself.[1] In Hong Kong, we had become socially acquainted with a few emigrants from Chen Village, and informal conversations had led to interviews, to introductions to former neighbors, and to further interviews. In all, in 1975–76, in 1978, and yet again in 1982, two dozen emigrants from Chen Village, including both peasants and urban-born young people who had settled in the village, shared their personal recollections with us. Of the 2,000-plus pages of interview transcripts that were obtained, almost 200 dealt with Chen Village's experiences with different remuneration systems.[2]

Cumulatively, the transcripts bring out clearly how each of these systems influenced work incentives, how each affected the ways in which the peasants interacted with each other, and how each system, by altering the perceptions of the villagers, influenced how they would react to the next system to be tried. It became a history of increasingly sophisticated mechanisms that elicited increasingly complex responses from the labor force, a history that culminated, by the mid-1970s, in a disillusioned rejection of the idealistically Maoist Dazhai system. It is a story that clarifies why, in recent years, China has chosen to decollectivize.

This chronicle of Chen Village's experiences with socialist wage schemes begins with the failure of the Great Leap Forward. Because of the Leap, the village had had to reerect a socialist economic structure beginning from square one.

## Starting Over: The Trauma of the Great Leap Forward

The Great Leap Forward had been economically catastrophic for Chen Village. In 1958–59, at the height of that confused utopian campaign, direct material incentives in Chen Village had gone by the board. Harvests had been siphoned off into an enormous pot the size of the entire local marketing district of nine villages (grandly dubbed a people's commune). The peasantry had been allowed, free of charge, to eat as much as they wanted in public mess halls, as an advance on future harvests. Peasants sarcastically recall it as the Eat-it-all-up Period: for when disorganized production led to bad harvests in 1959, the public granaries had already been emptied. The Chen Village peasants responded in their own best interests. Had they still obeyed the cadres and gone out to labor in the fields, their yields would have been diverted into that oversized communal pot from which other villages could draw. The Chen Villagers would have expended precious calories without any assurances of adequate returns. A hungry peasant's wisest strategy was to stay at home conserving energy, leaving the village fields untended. The Chen Villagers remained unwilling to put in any effort even when the state subsequently reduced the collective pot to the size of the village. A pot that fed close to a thousand neighbors still provided insufficient incentives to resume work.

To rescue rural socialism from the nightmare into which it had fallen, the government in 1961 was obliged to rebuild the agrarian economy from the ground up. A two-pronged program was enacted. As the first prong, each of Chen Village's five neighborhoods was organized into a "production team," to which was granted ownership and control over a fifth of the village's land. Team members were to share among themselves the harvest proceeds from this land. The idea was to create a collective unit small enough for them to perceive the relationship between their own contribution of labor, their team's productivity, and their own family's benefits from it. Indeed, to assure that the peasants would see these connections, each of the five neighborhoods was further divided half a year later to form even smaller teams, each containing some twenty to twenty-five families.

The teams were granted fairly wide decision-making powers. They were allowed to elect their own leadership and to manage most of their own affairs. They held the rights, for example, to determine how wage payments were to be arranged and who in the team would receive more and who less. The party bureaucracy, however, always retained certain important controls over the teams' plans. To fulfill national plans, it had the right to dictate every year how much acreage each of the Chen Village teams would have to plant in rice rather than in profitable vegetable crops; and it specified annually how much of this rice the teams would have to sell to the state. This system of production teams, the prerogatives granted to them, and the state's exactions of rice quotas have persisted down to the present day.

**Household Contracts**

A new remuneration system in 1961 comprised the second prong of the state's efforts to end the economic depression. Taking account of the Great Leap Forward's debacle and the peasantry's withdrawal from collective labor, the new incentives program had best be one that did not require much cooperation among households nor much supervision from cadres. Under Liu Shaoqi's direction, the government accordingly gave its blessings to a system called *baochan dao hu*. It literally meant "contracting production to the household." In Chen Village, lots were to be drawn each year and a portion of lowland rice paddy and a portion of hill land parceled out to each family. At the start of the season, the family would be provided with fertilizer and seed and given sole responsibility for planting and weeding its allotted fields. Each plot of land had a quota attached to it. A certain field carried a quota, say, of 300 pounds of rice, and at harvest time the family would have to hand in that amount of grain in exchange for workpoints, say 300 workpoints. The team would sell most of the harvested grain to the state and would disburse the proceeds, both in money and in kind, to each family in accordance with the numbers of workpoints it had amassed.

Initially, a family was allowed under the household contract system to keep for itself any grain it had harvested above its quota. But after about a year this rule changed. The government wanted to regain greater control over grain supplies. So the grain was to be harvested collectively, and all of it was to go to the team. But to keep the families working hard to produce more, the household and team were to share the proceeds from the team's grain sales in the following way. Quotas were again set for each field, with workpoint penalties imposed now for underquota harvests and a progressive workpoint bonus paid for all surpluses. An extra 150 pounds over the quota would, say, earn a family 200 extra workpoints.[3]

For the time being, the household contract system was popular. The Chen villagers once more were working productively and eating regularly, and they were thankful. But some families did far better than others. Peasants who were adept at agricultural planning and who had a number of capable teenagers available for labor profited most. They not only were able to earn those progressive bonuses; some of them even bid for extra plots of team land on which to raise vegetables and bid, too, for the special bonuses attached to tending the team's livestock. The household contracts did not particularly serve the interests, however, of families with children who were too young to work, or where the husband was weak or sickly or poor at agricultural planning. Since losses as well as gains were exaggerated due to the system of progressive penalties and rewards, these families were struggling desperately simply to avoid falling short of their quotas. As the depression receded, a constituency in the village was building against the contract system.

This constituency had the support of the state. The authorities had encouraged the contracting to families only as a temporary, marginally socialist expedient. In 1963, once the crisis caused by the Great Leap Forward had eased, very strong "suggestions" began flowing down from Beijing to move back toward a more collective system.

Though it was the poorer households in Chen Village that welcomed the government's demands, even the better-off families went along without any reported grousing. All of the peasants, the labor-strong families included, lived precariously close to the border of economic survival. No matter how strong and capable a particular husband and wife might be, the possibility of an infestation of their own small plots or a broken leg or unexpected illness posed ever-present threats. According to interviewees, these Chen Villagers, like peasants elsewhere, wanted more than just a chance to maximize their incomes; they wanted also to minimize risks.[4] A major attraction of socialism in the 1950s, interviewees report, had been its promise of greater financial security. A system of collective agriculture provided a peasant with the cushion of sharing in broader economic resources than he could manage on his own.

## Task Rates

Under the new remuneration scheme, team members would cultivate the team's fields together and share in the revenues generated. Called *baogong* (contracted work), the new arrangement employed task rates and piece rates to determine how big a slice from the collective pie each peasant earned. It was necessarily a complex program, because of the complexity of crop rotation in China. Unlike work on a factory production line, a peasant often had to complete a fair number of different farm tasks in the course of a day, and entirely different sets of tasks in different seasons.

Some of these tasks were much more highly rewarded than others. In particular, those associated with men were markedly better paid than those associated with women. Men and women normally worked separately. In the dry season, for instance, when dredging the nearby river the men were the ones who dug out the mud from the river bottom while the women hauled it up the river bank and packed it into the dikes. The men were paid for each bucket they filled and the women for each one they toted. It was the women's work that required the greater skill and effort, since the dikes were tricky to ascend under the swaying loads of dredged mud. But over the course of an hour, the men's digging paid almost twice as much as the women's carrying. The village women did not complain, however; they concurred that their own lower status justified lower pay.

Most of the hundreds of chores required the constant supervision of workpoint recorders, who had to jot down precisely how much each peasant accomplished. But in a few of the agricultural seasons ways were found to dispense with the services of these workpoint recorders.[5] For instance, peasants at harvest time worked in tight-knit squads of a dozen or so members, much as they had done even in traditional times. Without having to break their work rhythm, half of the squad members cut the crop; others would rush the sheaves to a small thresher at the side of the field; two men worked the hand threshing machine; and the two strongest men hustled the loads of grain into the village. Since the pace of the squad members' work was so closely interlinked, workpoints were awarded to the squad as a whole based on the tonnage harvested. In this "group task work" the members would hold a postharvest session to appraise each other's labor contributions and to determine among themselves how to divide up the totality of squad workpoints.

But most of the year the peasants worked for individual task-rate payments. On the whole, they were satisfied with this. But particular problems arose that invited complaints. For one thing, individual task rates necessitated reams of accounts; yet most villagers were illiterate and basically innumerate, and there were too few trustworthy personnel in each team capable of doing the daily accounting. Worse, team members would constantly bicker with the workpoint recorders about how many workpoints they deserved, and some also would squabble and jockey each morning to get assigned to tasks that either provided the most workpoints for a day's work or were easy in terms of the workpoints allotted.

More significantly, the task-rate system primarily paid people for quantity, at the expense of quality. During the planting season, for instance, task rates rewarded women (transplanting was women's work) in terms of the numbers of rows of rice seedlings each could transplant in the flooded paddy field. Hurrying to earn more, the women did not always bother to push the seedlings' roots in firmly. Some plants eventually would disengage and float to the surface. The very way the payments were computed encouraged a woman to keep her eyes fixed just on the size of her own slice of the collective pie, not on the productivity and income of the team as a whole. It ultimately led to suboptimal crop yields. Indeed, the task-rate system was less satisfactory in this respect than the household contract system, which had rewarded a peasant family only if it produced higher yields.

## Learning from Dazhai

In early 1965 a cadre work team entered Chen Village to push through a campaign variously called the Four Cleanups or the Socialist Education Campaign. This government work team, which stayed for almost two years in the village, was made up of rural officials and a couple of university students. It had come into the village to root out corruption among the local cadres, to revamp the methods of agricultural production, and to regenerate the peasantry's faith in the party and its ideology.[6] As part of these latter efforts, the Four Cleanups work team introduced to Chen Village in early 1966 the blueprints for a brand-new system of remuneration—the Dazhai system.

Dazhai was a village in the hills of north China that Mao and the party were promoting as the model for rural China. Among its signal achievements, according to the work team, was Dazhai's development of a new mutual-appraisal wage program. Its underlying idea was to structure remuneration in ways that induced people to concentrate their attention upon the gains that would accrue from a larger team pie. Such a proposal was potentially feasible because the teams were, after all, profit-sharing cooperatives. If the team pie expanded, each family's portion would grow; it would gain if its neighbors worked harder. The Dazhai system played to this point. Under Dazhai, the peasants would sit in judgment of each other at periodic team meetings to determine what each peasant's work was worth. Rather than providing direct monetary spurs, the Dazhai system would employ social pressures.

This system of mutual appraisals was supposed to provide built-in quality controls on labor. During the transplanting season, for example, team members would see it in their own interests to commend and reward most highly an effective balance between speed and careful planting. There promised to be other advantages, too. If wages were to be apportioned at periodic appraisal meetings, a team could do without workpoint recorders and complex bookkeeping. It could eliminate the daily wrangling to get better work assignments.

But the Dazhai system was not just supposed to be economically and administratively advantageous. The work team officials spoke of it as ideologically superior, and they denounced task rates as inimical to socialist ideals. The task-rate program, they said, had encouraged a selfish competitive concern to aggrandize one's own interests at the collective's expense. The system had daily corrupted the "proletarian" consciousness of the peasants. The Dazhai system would now give them the chance to prove and enhance their moral worth by pursuing the interests of the team as a whole.

The Dazhai method of mutual appraisals was merely one part of a larger package that the work team was introducing into the village. Daily study sessions were inaugurated in which the work team cadres and a group of urban-educated youths who had settled in the village taught Mao's quotes and incessantly impressed upon the peasantry the sanctity and relevance of the quotes. A perfervid atmosphere somewhat resembling a religious revivalist movement was whipped up—and put to the services of the new wage system. The rhetoric of the Mao study sessions and of a new village broadcasting system repeatedly intoned: fight personal selfishness, devote yourself to the collective. Villagers were supposed to concern themselves with the collective good even beyond the point where their own interests and the team's interests coincided. The politically moral man or woman was supposed to remain behind to finish up work in the dark even if no one were around to notice it. At the approach of a thunderstorm, a moral person was supposed to collect the team's property first and only then look after his or her own animals and grain.

Paradoxically, even though a peasant was not supposed to be thinking of personal gain, rewards for such selfless attitudes were built directly into the Dazhai wage system. In the mutual appraisals, not just a member's strength and accomplishments were to be evaluated, but also one's orientation toward the collective and one's willingness to spur others on and to serve as a quick pacesetter. In short, the workpoint ratings were supposed to be treated as the community's judgment on each person's moral attainments.

These judgments on wages were intended also to have an ethical import in a separate respect. Under the task rate system the strongest man in a team had been earning almost twice as much as the weakest man. An explicit goal of the Dazhai system was to narrow that gap. There was, again, a certain readiness within the village to pursue this. Many of the peasants, according to interviewees, agreed that the egalitarian leveling of the land reform and of the collectivization period of the 1950s had been justified.[7] The Dazhai system was presented as a further development of that tradition. It was supposed to reduce inequalities in earnings precisely because attitudes and not just strength would count. The weak man who tried his best would be given credit for it in his pay. The Four Cleanups work team thus particularly could turn for active support to all those villagers who would directly benefit from the Dazhai program: the peasants who were weaker or, more precisely, who came from labor-weak households.

On the other side, whatever their agreement as to the greater morality of the Dazhai system, villagers who foresaw that their household's interests might be damaged generally were hesitant to embark on the new program. As one of the rusticated urban youths recalls, ''Many of them thought that the task-rate system really wasn't so good; but since they themselves were good workers they wanted to keep it.'' Moreover, the new system entailed an untested risk; some of the team heads quietly voiced a concern that it would prove unmanageable and damage yields. They apparently felt that a community such as Chen Village, which was not far above the subsistence level of production, had precious little room for experiments. Had the choice been left in the hands of the teams, it seems doubtful that they would have ventured into the Dazhai system.

But in the midst of the Mao study campaign, no one was willing publicly to raise any self-interested arguments or conservative doubts about a program that so patently seemed progressive. There was a particular reluctance to object while the Four Cleanups work team was in charge of village affairs. Only recently the work team had investigated and browbeaten most of the peasant cadres for alleged acts of petty corruption. None now wanted to risk incensing the work team by casting doubts upon its new mission. All ten production teams ended up endorsing the Dazhai experiment without serious public debate.

## Operating Dazhai-Style: The First Years

The work team proceeded cautiously, knowing that the Dazhai method would not be easy to run effectively. One of the ten teams—a team which had a very capable leadership and seemed the least divisive internally—was asked to pioneer the program with the work team's aid. In the time-tested manner of campaigns, this team served as the exemplar for the other nine teams. One of the members of that team recalls, ''The team heads and team committees and Mao study counsellors from all the production teams came to observe how our team did it. So did all of the village's party and Youth League members. That way, when they saw how our team could handle it, they wouldn't feel that it couldn't be done.''

The onlookers went away largely convinced. It seems from interviews that most of the villagers wanted to believe in the new system's efficacy. Since their teams were already committed to the Dazhai program, they hoped it would confirm the promises made in the Mao study sessions. Some of them wanted the opportunity to prove their moral superiority; and all wanted to achieve improved living standards. If this new method could spur better work and higher production, they were willing to give it their best try. The Dazhai system, by its very nature, relied upon such a willingness. As one of the peasants remarks, ''In the beginning, *because* folks had trust in the Dazhai system, things went well in our village.''

The success of early workpoint appraisal sessions depended not just on this spirit but also on careful planning. Before each of the appraisal sessions, the team head would gather together several strong, politically activist men to rehearse a smooth start to the session. It would be arranged for a few of them to volunteer from the audience to appraise themselves first, both to break the ice and to impart a proper atmosphere to the meeting. This was important because each person's appraisal was to start with a self-evaluation, something many peasants felt awkward to attempt in public.

At the main session, one of the prearranged volunteers would rise and deliver a short self-deprecating speech, to the effect that he had been making an attempt to follow the teachings of Chairman Mao, but to tell the truth, on such and such occasions he had not worked hard enough. He thought he deserved only 8.5 workpoints a day. He would try much harder in the future to work selflessly in the masses' interest.

Such a self-appraisal inevitably involved play-acting. There are a great many formal sessions in China in which people use an official rhetoric to say exactly what they are expected to say. They were now learning in Chen Village how to do so with self-evaluations. But in this initial phase of Chen Village's experience with the Dazhai system, say interviewees, the rhetoric was also taken seriously by many of the speakers. They did not want to seem hypocritical to themselves and others. Many of them did want to live up to Chairman Mao's teachings and do better work. When necessary, they did work overtime on their own. Moreover, such team members generally were determined that others should equally live up to their promises. A former Mao study counsellor remarks, ''People didn't dare make hypocritical promises at the meetings, because they'd get criticized if they didn't indeed do better. In the countryside this was easy to see.''

After a person had presented his or her self-appraisal, other teammates would be asked to add their evaluations. But in those first months ''they felt uncomfortable,'' recalls one of the villagers. ''They'd never before judged other people right to their faces in this way. And besides, they felt that if they dared to speak out bluntly, when their own turn came people might raise lots of opinions about them.'' Those willing to speak up were usually the people who were beyond criticism—a few of the best male workers. For someone like that self-effacing activist who had requested 8.5 workpoints, they would offer praise and a suggestion that he receive a full 10 points a day—the top of the wage-point scale. One teammate after another would be called upon to concur.

At these early meetings, recollects an interviewee, not a single peasant in his team dared to request the full 10 workpoints. Almost all had requested less than they deserved and were subsequently upgraded by teammates. But all those who regularly had come late to work, or had lagged behind in their labor, or had missed team meetings came under criticism and saw the disapproval reflected in their workpoint rating.[8]

Every day, moreover, the team leaders and the strongest and most committed of the team members were organized to labor at a faster-than-usual clip that forced others to step up their own pace. Recalls the former Mao study counsellor:

> These activists would be working very hard and would be mad others weren't trying so hard. [When the labor squad had its lunchtime Mao study session] they'd take the lead in speaking up about these things. They wouldn't mention any names. If you wanted to make a self-confession you decided on your own. But if they'd bring up a problem and you didn't correct yourself, then later you might get mentioned by name. "How come you aren't doing it right? We studied about this in the afternoon and you're still doing a bad job! Do you want your team to have enough to eat or not?" Ha, if you had these activists on the scene, you weren't going to have any laziness.

In these circumstances, the Chen Village teams did achieve an upward surge in production. The work team had introduced a Green Revolution hybrid rice strain. To grow properly, it required better water control and heavy amounts of fertilizers. New irrigation ditches needed to be dug; the paddy fields needed to be leveled; large quantities of extra compost needed to be collected. Under the Dazhai system, both the quantity and quality of the peasants' work rose more than enough to meet the new labor demands. The Chen Villagers enjoyed the payoffs: by 1967, rice yields had nearly doubled. These successes boosted peasant morale and helped sustain intact the social pressures necessary for the Dazhai method's smooth operation.

## A House Divided Against Itself

The Dazhai program was a double-edged sword, however, whose methods would, over time, cut both ways. On the one side, the system succeeded partly because the teams were communities whose members worried about each other's opinions. On the other side, the Dazhai program's eventual undoing was precisely this: it encouraged the peasants to be too sensitive about their standing among their neighbors.

What created particular problems was the Dazhai system's concentration upon attitudes. Any judgments of a teammate's attitudes necessarily were subjective, in ways that observable and quantifiable criteria like strength or speed were not. This was one respect in which the Dazhai system was much more difficult to manage than that earlier group method of harvest-squad appraisals. Appraisals then had been confined to how much each member had produced; and since a man or woman could not help being weak or elderly, a low income under that earlier system had not impinged directly upon a person's sense of integrity. But now, under Dazhai, much more than just payments for work done had become involved in workpoint appraisals. The judgment on overall performance and attitudes was perceived as a measure of what the community thought of a member as a complete person. A lower appraisal implied a lower status. Indeed, the imprecise standard of attitudes inadvertently became a means to downgrade the ratings of people who were cantankerous and generally disliked. To retain their standing even more than to retain their earnings, such people learned to shed their reserve and to begin arguing vociferously for the workpoint rating they felt they deserved.

To reduce the acrimony, little more than a year after the introduction of the Dazhai system the team heads quietly abandoned attitudes as a criterion for the ratings. They never officially announced it; to follow the right political line, the fiction that good attitudes were being rewarded had to be preserved. The team heads simply steered appraisals toward the actual work accomplished.

Yet the mold had been set. Villagers had become accustomed to viewing the ratings as a measure of their comparative status, and most remained competitively alert to what they and their neighbors received. Many were intent to receive as good or better a rating than teammates they felt were of the same level, and at the same time they tried to prevent those who had been appraised lower than themselves from climbing up.

Within a couple of years, only the most politically devout team members any longer made a self-deprecating little speech or requested undervalued ratings. One interviewee who idealistically insisted on doing so discovered, to her considerable dismay, that two women with whom she was at odds quickly took the opportunity to congratulate her on her honesty, and that bemused teammates declined to speak up to rescue her from her own too-modest rating. The next session she requested exactly what she considered she was worth.

A contrary manner of self-appraisal became increasingly common. Some participants began inflating their own appraisal in the hope that no one else would risk a quarrel by objecting. Rarely, though, did such a strategem work: some teammate or other would insist upon keeping their own workpoint values higher than yours. Arguments gradually became rife. A wage system that specifically was intended to reinforce cooperation among teammates was becoming the cause of growing contention. The appraisal sessions increasingly were pitching personal interests and egos against the team members' common interests.

The weakest members of a team, the ill, the elderly, and the handi-capped, did not tend to be among the most quarrelsome at these sessions. They were embarrassed that they were a burden on the team and avoided making any scenes. Nor was it the strongest and most capable men who normally agrued for better points. Their status was unchallengeable; they felt no need to quibble over a tenth of a workpoint. That comprised, after all, only 1 percent of their earnings, just a few cents per week. A tenth of a workpoint—the likely issue in disputes—only mattered to those concerned with fine distinctions in status. Almost always, agree interviewees, these disputants were about average in their abilities. They felt they were making a real contribution to the team but some of them obviously felt insecure about their status and were jealous to safeguard it.

Interviewees agree, too, that teenagers tended to be a focus of the rating disputes. A teenager's physique and strength could show marked development from one agricultural season to the next, and their ratings therefore required constant reevaluation. They normally were too embarrassed to speak up in their own behalf; but to their mortification, their mothers often felt no reluctance to pick up the gauntlet for them.

Indeed, almost every interviewee claims that, to a very noticeable degree, women were more argumentative in these appraisal meetings than their husbands. Two reasons were discernible. First, as explained by a young woman from the village, men had much greater opportunity to achieve a genuine status of their own. A majority of the men at one time or another would hold a post at the team or village level, but even the most capable and ambitious women were boxed in as women. The *symbol* of status represented by the workpoint system accordingly became more important to them. Second, being blocked from personal achievements, women more than men saw their social standing in the village as intertwined with that of their family and kin. Hence, they tended more than men to intervene in disputes to support the claims of relatives, thereby prolonging and widening the quarrels.

These difficulties were compounded in 1969 when the commune administration succeeded in forcing reamalgamations of the very same teams that in 1962 had been divided in half. The new, larger teams would be able to control irrigation better and mechanize more efficiently. But with a doubling of each team's membership, the appraisal meetings became even more unwieldy. There were now twice as many teammates of one's own age and own capability with whom to compete, and more kin in the same team whom one could turn to for support. The government's national push in the late 1960s and early 1970s to combine teams inadvertently was undermining the government's simultaneous efforts to preserve the Dazhai system.

As the appraisal sessions in Chen Village grew more acrimonious, the best and most prestigious workers no longer were so willing to offer their disinterested appraisals. They did not want to risk getting unnecessarily caught up in a feud. The team heads themselves had to step in more regularly as the final arbiters of ratings. They began to dread the sessions.[9] Team members who received less than they had asked for carried their resentments out into the paddy fields; and the daily work of the cadres was made all the more difficult. The Dazhai method increasingly was generating what the Chinese call ''mass/leadership contradictions.''

Team cadres took the easy way out—scheduling fewer appraisal meetings. In 1966, when the system was first introduced, the meetings had been held every two weeks. According to government suggestions, once the system was successfully established fewer meetings would be needed; but Chen Village's cadres pushed for fewer sessions than the government could ever have expected. By 1971, the appraisal meetings were being convened only once every half year (at the very least, grain and cash had to be distributed after each of the two harvests).

But this avoidance of sessions only heightened the tensions when the peasants met. A half year's face was now at stake. Moreover, even the miniscule difference of a tenth of a workpoint per day now counted for something in monetary terms. Peasants who earlier had stayed out of the arguments found it worthwhile to expend that extra half hour at a meeting angling for a better rating. The sessions were becoming impossible to keep under control. Whereas the initial appraisal meetings of 1966 and 1967 had taken a couple of hours to settle, by 1970 the meetings frequently were lasting till dawn—and sometimes had to be resumed the next evening and occasionally even a third and fourth night. Each time the teams were left physically exhausted and internally divided.

By 1971 the team heads had concluded that the solution lay in simply giving each team member exactly the same rating as in the previous meeting. It seemed far better to have a brief meeting without appraisals than to endure the interminable arguments and subsequent backbiting. But by unofficially converting the Dazhai system into a simple system of fixed salaries, the team heads were now granting people workpoints regardless of their work performance.[10]

The Dazhai program had entered a final phase in its decline. Up to now, whatever the terrible quarrelling at appraisal sessions, the Dazhai method had operated reasonably well out in the fields. Teammates had continued to pressure each other into hard work. Some had been motivated by the prospects of higher yields and improved living standards all-round. Others, less public spirited, had been angling to prove themselves better than their rivals. Yet others had been spurred by their fear of scrutiny and criticism. But the Dazhai system itself had now altered. Having evolved into fixed wages, even the incentives to compete for a higher status rating had been removed. Just as troubling, as far as the best workers were concerned, there no longer was even a fair spread in workpoints between the best and worst laborers.

Through the years, a group dynamic had been at work in the appraisal sessions. Whenever quarrels had arisen between workers of equivalent abilities, the teams had found that these could be resolved most conveniently by lifting the workpoint rating of the lower disputant to the same level as the higher, rather than infuriate the higher by lowering his or her points a notch. These inflationary compromises resulted eventually in a very narrow workpoint spread between the best and the worst, but any across-the-board attempt to rewiden the wage gap would have encountered the hostility of the many households that had benefited. The teams could not afford any additional divisiveness, and so cadres ignored the problem. In 1969, when a new cadre work team stayed briefly in Chen Village, it expressed concern that the workpoint gap no longer provided equitable economic incentives, and the team heads concurred. But when the work team left, not one of the teams tampered with the narrow spread.

Just how narrow can be seen in the following figures. At the initial 1966 appraisal meetings, the men's apportionments in one particular team had ranged from a low of 7.3 points to a high of 10, and the women's from a low of 5 points to a high of 7.5. By the early 1970s, the best still received 10, but the average man was getting 9.5 and the weakest 9.[11] The best male worker was now earning only a slim 5 percent more than the team's average and only 10 percent more than his least capable and energetic neighbor. The spread among the women had become even narrower, from a high of 7.5 points to a low of 7.1.[12]

When production annually had been expanding, the strongest peasants had not particularly minded that their less productive neighbors were obtaining a disproportionate share through this gradual narrowing of workpoint differences. Since their own family was living better than ever before, they felt they could afford a measure of altruism: "They didn't want to hurt other people, so they felt 'Well, it doesn't really matter.'" But by about 1970, the village's economy started souring.

Over the years the Green Revolution grain hybrids had been growing less resistant to blight, and the state was not providing substitute hybrids of any quality. But a more potent factor in the village's economic troubles was, simply and purely, the heavy hand of China's bureaucracy. The national, provincial, and county bureaucrats, having recently been pummelled in the turmoil of the Cultural Revolution, seemed concerned primarily with their own political skins, and in the uncertain political winds of the early 1970s they were pushing any policy that seemed safely leftist. They were pressing the village in loyalty campaigns to contribute extra grain to the state at below-normal prices; to "self-reliantly" grow crops such as wheat and cotton that were woefully unsuited to the climate; to forgo profitable team vegetable plots and to fill up money-making fish ponds in order to plant yet more grain; and then, as national slogans and policies periodically flip-flopped, to reexcavate fish ponds and "diversify."[13] The Chen Village peasants frustratedly watched the value of their workpoints decline from one year to the next. This decline can be measured through a year-by-year accounting of what the best male workers from one of the village's richer teams could earn in a day's work on the team's fields (see table 5.1).

The stronger team members, finding themselves with a shrunken portion of an annually shrinking pie, began to feel that neighbors who contributed less to team output were unjustly benefiting at their expense. Some of them began to slacken in their work and to recoup their declining earnings by putting more energy into their private plots. As increasing numbers of the best laborers stopped serving as pacesetters, the pressures upon the others to keep up were relaxed. Ultimately whole teams began to slough off. The influence of the peer group milieu had swung almost 180 degrees. Before, even lazy members had had to work well lest they be accused of taking advantage of all their neighbors' hard work. Now, conversely, people were shunning hard work for fear of being taken advantage of.

Table 5.1

**Daily Workpoints Earned by Best Male Workers, 1964–1977 (in yuan)**

| Year | Workpoints | Year | Workpoints |
|------|-----------|------|-----------|
| 1964 | 0.50 | 1971 | 1.00* |
| 1965 | 0.80 | 1972 | 0.90 |
| 1966 | 0.90 | 1973 | 0.80 |
| 1967 | 1.00 | 1974 | 0.80 |
| 1968 | 1.15 | 1975 | 0.45** |
| 1969 | 1.10 | 1976 | 0.80 |
| 1970 | 1.00 | 1977 | 0.70 |

*Notes*: *In 1971, Chen Village's richest team could provide its best workers a daily pay of about 1.10 yuan, while the poorest team could offer only some .60–.70 yuan, a difference greater than the difference in earnings between the highest- and lowest-paid members of the same team.
**1975 was a year of very adverse weather. It was the third wettest in the 130-year history of Hong Kong, less than 100 miles away, and the winter was Hong Kong's coldest on record.

The Dazhai system was malfunctioning in villages throughout China. For once, the party leadership in Beijing grew concerned. Hints appeared in the government press as early as 1971–72 that, if need be, teams should abandon the remnants of their Dazhai effort and revert to task rates. Many of the team cadres in Chen Village would have liked to comply. But they were reluctant to go out on a limb. Until recently, task-rate systems had been condemned in the mass media for encouraging selfishness and "petty-bourgeois thinking." More than mere hints would be required now to dispel the team cadres' fear of plunging into a dangerous political blunder.

Higher levels in Guangdong provided them in 1973 with an "emperor's clothes" solution. A new slogan was announced: "Repudiate Liu Shaoqi's task rates; permit Mao Zedong's task rates." So far as the Chen villagers could see, the two systems seemed entirely similar; but the verbal legerdemain made it safer to revert.[14]

Under the Dazhai system of face-to-face appraisals, the stronger workers had felt embarrassed to recommend sharply reduced earnings for weaker friends. Under task rates, however, everyone would simply be paid in accordance with what they produced. There would be no need ever to confront neighbors directly and personally in any meetings. The weak would be able to blame not neighbors' appraisals but only their own slow labor for their low incomes. Task rates, in short, provided an effective means to rewiden the earnings gap between the stronger families and the weaker.

The team memberships accordingly were divided on the issue of Dazhai versus task rates. An interviewee recalls,

> Our labor squad's head is strong and so's his wife, So they wanted to revert to task rates. But the squad head's elder brother is lame and his wife is near blind, and they *hated* task rates. So right among those relatives, you found a "contradiction," one wanting to go back to task rates and the other strongly opposed.

A couple of the village leaders and a number of the rusticated urban youths sided with the weaker and the poorer. But almost all the team heads threw their weight toward task rates. It would bring an end to their unhappy need to badger team members into doing their work. One after another in the summer of 1973, the team heads steered their teams back into task-rate programs.

## Task Rates and Contracting Revisited

Even though task-rate payments again were structured exactly as under "Liu Shaoqi's task rates," in one respect the revived system now did operate differently. In the mid-1960s, before Dazhai, few ordinary peasants would have bawled out a teammate who did sloppy work, for traditionally to cause someone to lose face had been a breach of etiquette. But under Dazhai, a "social contract" had been erected. Under its ground rules, anyone overly careless could be loudly reprimanded. This Dazhai norm now carried over into the revived task-rate system's operations: "Nowadays [1975], if folks get caught transplanting the seedlings too carelessly, just so as to get more planted in a hurry, other folks would tell them off." The concern about "face" had been turned on its head: whereas earlier the norms involving face had protected careless workers from being criticized, the threat of public embarrassment now helped keep potential offenders in line. But even with this, the renewed system all but invited shortcuts. Interviewees report that when task rates again became the basis for rewards, the quality of work worsened.

Moreover, even the revival of task rates in 1973 did not result in appreciably harder work. With agricultural profits declining, most peasants still were finding it to their individual advantage to put more of their energies into their higher-paying private endeavors. Indeed, the task-rate system gave them greater opportunities than the Dazhai system to take off early from work to do so.

A great many of China's villages, by the accounts of China's official news media, were caught similarly in agricultural slumps. Large numbers of these villages apparently were facing considerably greater problems than Chen Village. The party Central Committee in Beijing reacted in the late 1970s. Mao had died; the radical Gang of Four had fallen shortly thereafter; and the new leadership, once entrenched in power, felt willing to countenance a markedly further retreat from the Dazhai system. If under task rates China's peasants were not working hard enough or effectively enough, or were not caring enough about the quality of their work, then the teams ought to revert to a system that tied payments even more directly to the productivity of each person's labor. It was now felt that perhaps the very size of the teams (that large pool of some fifteen to fifty households) inhibited peasants from seeing clearly enough the connections between their own work contribution and their returns from the collective productivity. The national leadership decided that the solution would be to encourage teams to decentralize and to hand over decision making and the sharing of profits to smaller labor groupings.

The new directives came down through the Guangdong provincial party in early 1979.[15] In compliance, the Chen Village teams organized small labor squads of "several families," which negotiated quotas with the team leaders, planted and tended the fields, and divided among themselves the quota payments and the bonuses for surplus production. Within another year, again under official urgings, Chen Village shifted to individual household contracts, paralleling the contract system of the early 1960s. Indeed, production was allowed to decentralize to the point that even harvesting was entrusted to family hands. Beyond a quota of grain that had to be sold to the state, households were to be allowed to keep or sell privately all the crops they grew. To encourage them to improve soil quality, it was even tacitly agreed by 1982 that fields would remain in the same family's care fairly permanently.

Emigrants from Chen Village relate that almost all of the peasantry approve of the new system. The recent bad experiences with collective agriculture—the failings of the Dazhai system, the party officialdom's blundering economic demands of the 1970s, and the persistent slide in workpoint values—had soured most of them on collective production. As much to the point, to work independent small holdings no longer seemed so precarious a venture to Chen villagers. Earlier, in 1961–63, when they had been living close to the margins of economic survival, the risks entailed in the household contract system had worried them. But now new economic connections with Hong Kong were being cemented which sharply reduced their risks. In early 1980 the government announced that all of the villages in the county would be allowed to sell their agricultural produce directly to Hong Kong. The door had been opened to very high and relatively secure profits in vegetable truck farming. Whereas some of the peasantry elsewhere in China may have felt wary about risking a system of contracted fields, few Chen Village peasants any longer saw reasons to object.

## Conclusions

Chen Village, in a period of two decades, had swung from a household contract system to the Maoist experiment of Dazhai appraisals and then, like a pendulum that has mounted the crest of its arc, had fallen back by degrees to the opposite crest of the arc and a revived program of contracts. By 1981, agricultural production, though not land ownership, had been largely transferred into private hands in the village.

What should we make of this long swing into the Dazhai system and back?

It is clear, for one thing, that in Chen Village there was a pattern to these shifts. The household contracts of the early 1960s, the subsequent system of task rates, the Dazhai method, and the reverse shifts were all introduced partly as solutions to the problems that had emerged in the immediately preceding programs of remuneration.

What we witnessed, however, was not simply an evolving sequence of local problems and local responses. Intervening was a complex interplay between state and collective. Each new program was "suggested" by the state. This intervention was based partly upon the state's general awareness of the problems faced in China's collectives, but sometimes, as with the Dazhai system, it was based also upon the party leadership's ideological commitment. Once a program was introduced, the teams did possess some leeway in modifying each remuneration system as they wanted, but they always had to keep an eye fixed on party policies to make sure they were operating within the parameters of what was politically permissible. For instance, the Chen Village teams were able to alter the shape of the Dazhai program several times as stopgap remedies to local difficulties, but they had to stick with an ostensibly Dazhai-type framework until new upper-level suggestions let them pull free in 1973.

It was observed that in Chen Village the Dazhai program operated reasonably well in the rice fields (if not in appraisal sessions) until about 1971—and within just another two years the village was allowed to abandon the Dazhai scheme. But that had not been the experience of many other communities (including several in my small sample of other Guangdong villages). Caught from the start with a Dazhai system they could not handle successfully, they still had to make the pretense of an effort, at a considerable cost to agricultural productivity.

In this respect, despite all the problems it faced, Chen Village was among the more fortunate villages. In fact, its experiences with the Dazhai system were considerably better than any of the other villages that I know about through interviews. A Four Cleanups cadre work team carefully had laid the groundwork in Chen Village for the Dazhai experiment; the new system had been willingly accepted by many of the villagers; and their faith was rewarded for several years with rapidly rising living standards. Chen Village accorded us a glimpse of a community grappling, initially with success, to operate almost solely on the basis of collective interests.

This being the case, what do Chen Village's experiences say about a production team as a "community," or about its members' capacity to cooperate for collective goals?

It was observed, for a start, that the Chen Villagers, though always concerned for their own family interests, were not exclusively so. They understood that concerted efforts to raise agricultural productivity would raise living standards all around. And most of them even were willing to support the Dazhai method's more egalitarian apportionment of wages—just so long as the increased welfare of their less capable neighbors was not ultimately at their own family's expense.

The difficulty was to reconcile the family's and the community's interests. Here, the production teams always faced the "free rider" problem. Not all of one's neighbors could be expected to put collective goals on a par with their own more narrow interests; and thus, not all of them necessarily would contribute their fair share of work. That being so, who but a fool would exhaust himself laboring hard in behalf of such "free riding" neighbors?

The Dazhai experiment promised answers to this conundrum. As observed, it was introduced into Chen Village in the midst of a major Mao study campaign. With Mao's sacred thought preaching the higher morality of the collective road to prosperity, shirkers under Dazhai risked being tagged as politically and morally backward. Pressures could effectively be brought to bear on this point, for the peasants' private concerns had always been of two types. While concerned with their own economic well-being, many team members were equally concerned about their standing in the community. In the mutual-appraisal sessions, this desire for a better status could be employed to counter the tugs both of laziness and of personal short-term interests.

There were dilemmas here, however. To ensure conformity, the Dazhai system would play upon the strength of community sanctions— but only by exacerbating the small-minded anxieties to retain "face" traditionally common to an in-grown village milieu. In the end, as has been seen, the Dazhai wage program bent and broke under tensions partly of its own making. When the heightened competitive desires among neighbors to defend their status resulted, at one and the same time, in an erosion of team morale and an inordinately narrow wage gap, the strongest and most respected workers began to consider themselves the victims of free riders. As has been observed, when they opted to spend more of their time and energy on their private plots, the Dazhai system of work incentives and community sanctions collapsed from within.

Ultimately, in that long reverse swing of the pendulum away from the Dazhai system, the household contract system resolved the same problems of controlling free riders that the Dazhai method had tackled—but from the opposite extreme. Household contracts eliminated the difficulties associated with collective cooperation by, quite simply, terminating collective cooperation.

By pushing for collective goals by way of instigating a contest among team members to preserve "face," the Maoist Dazhai program had frayed the very social fabric it sought to strengthen. In Chen Village, it had left in its wake a diminished faith in collective solutions—indeed a willingness among the peasantry to dismantle the collective organization altogether. The failed Maoist dream has bequeathed an ironic legacy.

## Notes

1. See Anita Chan, Richard Madsen, and Jonathan Unger, *Chen Village: The Recent History of a Peasant Community in Mao's China* (Berkeley: University of California Press, 1984). Thanks are due to my fellow authors for permitting me to use collective research data for this chapter.

2. Information that I separately gathered from emigrants from four other Guangdong villages suggests significant differences as well as similarities in different villages' reactions to wage systems. An excellent discussion of this, using a survey questionnaire covering sixty-three Guangdong villages, is included in William Parish and Martin K. Whyte, *Village and Family in Contemporary China* (Chicago: University of Chicago Press, 1978), pp. 62–71.

3. Teams selected their own method for contracting production. Other localities

in the same region employed slightly different schemes. On this, see *Southern Daily* (Canton), June 21, 1962, p. 1, translated in *Union Research Service* (Hong Kong: Union Research Institute), vol. 29, p. 323. Also see the party documents from Baoan county, Guangdong, in *Union Research Service,* vol. 27, pp. 115, 124, 137–38, 142, and 151. Even earlier, in the mid 1950s, village cadres in many parts of China surreptitiously had resorted to contracting fields to the households in order to relieve the difficulties of coordinating and supervising collective cultivation. F. W. Crook, "Chinese Communist Agricultural Incentive Systems and the Labor Productive Contracts to the Households, 1956–65," *Asian Survey* 13 (May 1973).

4. This type of calculation has tended to be true of peasants the world over: "Living close to the subsistence margin and subject to the vagaries of weather and the claims of outsiders, . . . the peasant cultivator seeks to avoid the failure that will ruin him rather than attempting a big, but risky, killing," James C. Scott, *The Moral Economy of the Peasant* (New Haven: Yale University Press, 1976), p. 4.

5. During the postharvest season, the village might engage in road construction, and here each person was given a fixed length of road to dig for a fixed number of workpoints. As the reward for hard work, those who finished fastest were able to leave early to work on their private plots. Or, if there were no public project to finish and little need to push team members to labor efficiently, the strongest workers simply would be designated as grade A and would receive 10 workpoints per day, grade B workers would receive a flat 9 workpoints each day, grade C workers 8 points, and men who were grade D 7 points, irrespective of how much they accomplished. With no monetary incentives to work hard, people would take it easy, resting from the strains of the peak season.

6. For a detailed description, see *Chen Village,* chs. 2 and 3.

7. Even a landlord's son, whose family had lost everything due to the new order, remarked to us that "the good side of socialism is that earnings are more equal so that even the lame and the weak survive."

8. The scenario of these initial meetings was not successfully followed in the other four villages in my sample. The principal reason may have been that a Four Cleanups work team had not played a hand in establishing the Dazhai method in these villages, nor organized a Mao study program to give the method the needed ideological underpinnings. In one of these other villages, "even in the very first evaluation meeting the people didn't really take political performance into consideration. When some did, others said, 'Are activist attitudes edible?' People would say, 'Counting task rates I'll work my ass off, but counting time, under this Dazhai stuff, I'll take the opportunity to have a rest.' . . . So both systems—task rates and time rates—were used simultaneously in my village without taking attitudes into consideration at all."

Martin K. Whyte has gathered descriptions from a number of interviewees from villages that similarly had set up the Dazhai program without the aid of a work team and which, from the start, never got the system to operate properly. "The Tachai Brigade and Incentives for the Peasant," *Current Scene* 7, 16 (August 15, 1969).

9. An emigrant from another village recalls a somewhat different scenario. In that village the peasants feared getting on the wrong side of the team cadres and their families; and the team officers eventually controlled the sessions for their own benefits: "At the early sessions, the most difficult to appraise was the team head's wife. She was lazy but had the nerve to get up and say she ought to be considered top grade. For a whole hour, there was silence; no one stood up to support her, no one to oppose her. And the team head himself wouldn't step up and say anything once she'd made such claims. Eventually she got awarded a high rating. . . . My main complaints with Dazhai were that the cadres were better able to take advantage of it than task rates, and that with Dazhai there was the feeling that if someone didn't work hard yet finagled high workpoints, then I won't work hard either."

10. At a different village in my sample, the team leaders went a step further. They

not only instituted a flat time wage; they entirely abandoned workpoint differentials and thereby removed the divisive associations between workpoints and status. "The highest—and the lowest—for a man was 10 points. All the women earned 9 points." Question: Why were there no differences at all in the wages of the men? "Because that would create conflicts."

11. The elderly, semi-retired men who did odd chores for the team received 8.5 points, while the old women who had not yet retired were given 6.

12. During the harvest seasons, when the teams were very short of labor, the men's top rating temporarily was set at 15 workpoints and the top women's at 12, to take account of the longer and tougher workdays. But the wage spread among the men and, separately, among the women remained as narrow as during the regular seasons.

It is evident in the figures presented in the main text that the lowest rating ever given to a man was higher than the higest rating allowed for a woman. The traditional view that defined all men's work as *ipso facto* superior to women's work had persisted. But in about 1968–69, this came under challenge. Some of the young women from Canton who had settled in Chen Village began agitating for better ratings for women and were able to convince some of the strong young unmarried women to join in. Recalls a young man:

> At that time other communes were talking about "equal work, equal pay" for both men and women, so the young women settlers felt they could begin talking about it in Chen Village. A woman of 7.5 workpoints actually might work better than a man of 9; but the problem was that the women couldn't plough. This was the weak men's last resort. Ploughing had always been a man's job, and it was an important job. There were two women in my team who could plough pretty well on their private plots, but the men would never ask a woman to do any ploughing when allocating labor.

As a group, the men of Chen Village were adamant: 7.5 remained the women's maximum.

13. See *Chen Village*, ch. 9.

14. In a village in a different region of Guangdong, according to an emigrant, a different political ploy was offered: "When Lin Biao fell, this so-called Dazhai system was said to have been distorted by him, as an example of his empty-headed politics. They began talking about adopting the correct policy of 'to each according to his work.'" That village returned to the task-rate structure in 1972.

15. Canton Radio, January 21, 1979, in *Foreign Broadcast Information Service: China Daily Report*, January 24, 1979, p. H1.

# 6

## Peasants, Ideology, and New Incentive Systems: Jiangsu Province, 1978-1981

## David Zweig

### Introduction

When in 1978 China's leaders called for reform in the rural incentive structures and the institution of a "responsibility system" (*shengchan zhiren zhi*) for agriculture, few analysts or, for that matter, Chinese could have predicted what the outcome of those changes would be. The original policy aimed at establishing a wide variety of organizational and remunerative techniques that linked the income of the individual peasant and his family to the quality or quantity of the agricultural goods produced. Yet, by 1983, the decollectivization of rural China had been completed and the individual household had again reemerged as the primary economic unit in rural China.[1] How did this transition occur? Although the case can be made that some localities were forced by higher level bureaucrats to decollectivize against their will, other areas responded to the positive economic opportunities offered by the new policies.

William Parish's introductory essay in this volume has already reviewed the development of this policy at the national level and provided a typology for understanding the multiplicity of incentives that have been acceptable at various times. This chapter analyzes the introduction and implementation of these new policies in three brigades in Jiangsu province between 1978 and mid-1981.

Three interrelated points form the central issues here. First is the relationship among freedom of choice, bureaucratic coercion, and economic rationality. Did the local units studied here adopt these tech-

My thanks to Martin Whyte, Penny Prime, Kevin Kramer, David Shambaugh, Michel Oksenberg, and especially William Parish for their helpful comments on an earlier draft.

niques for increasing labor accountability on their own initiative or under bureaucratic pressure? If the impetus for change came from the bottom up, peasants and local leaders[2] must have recognized the economic rationality of the policy before the central leaders attained consensus and officially sanctioned it. If implementation was due to pressure from above, middle-level bureaucrats either enforced compliance because local officials underestimated the policy's benefits or imposed conformity for its own sake, irrespective of the policy's utility to the specific locality. And if the latter situation is true, China has again failed to learn from past errors.[3]

The second issue, the nature of the Chinese peasant, also relates to the proper locus for decision making for this policy. Peasants respond to the world based on their own naked economic interests.[4] Local leaders, therefore, by consulting with peasants, should have been best situated to determine which policies would maximize economic production, which is the goal of China's central leadership. And due to different natural conditions and levels of economic development, localities should have responded in dissimilar ways to alterations in the collective system.

Finally, the leaders of the Chinese Communist Party (CCP) never really succeeded in resolving the conflict among peasant households, the collective, and the needs of the state. Rather than strike a balance between the three, the leadership has chosen instead to remove one actor, the collective, from the equation. But as the state permitted individual interest to play a greater role in the rural economy, how did the peasants respond to increased freedom? Did they support the collective or did their private interest propel them to dismantle the collective? Although today we know the result of the reforms, this brief glimpse of rural China in transition may help us understand how peasants and cadres would have responded to these incentives had they been allowed to determine the outcomes that best suited their local interests.

## The Local Setting

From January through July 1981 I carried out field research in three different areas of Jiangsu province. I talked with local leaders and peasants about these methods for increasing labor accountability and their attitudes toward them. Though all three sites were in the greater Nanjing municipality, they differ enough in leadership, wealth, terrain, and cropping patterns that comparing their different responses to various incentive systems provides valuable clues as to the appeal and liabilities of these systems throughout China. Table 6.1 sets out the major differences between the three units.

The decisions in these units concerning these incentive systems were based on political, psychological, environmental, economic, and technical factors. Politically, the attitudes of bureaucratic superiors and the authority local leaders had within their own unit shaped the adoption of new incentive systems. By rearranging labor organization and remuneration in the countryside, these accountability systems involved questions that had long generated political conflict. For local leaders these risky political changes were less dangerous when their bureaucratic superiors supported the policy shifts, and vice versa. Within each team or brigade, strong leaders imposed their own will, while where the brigade leaders were weak, more variety in implementation existed among the teams.

Psychologically, some peasants, fearing further flip-flops in agricultural policy, hesitated to invest energy in land that could have suddenly reverted to the collective. And local cadres, remembering the 1960s and '70s when political rather than economic mistakes threatened their careers, decided that being "left" was safer than being "right" (*ning zuo wu you*). From a Maoist perspective the stress on material incentives was clearly "rightist," so local cadres had to wonder if today's implementation could become tomorrow's "rightist" error. Given the peasants' and cadres' fears, some localities did not accept the new policy until authoritative statements emanated from the Central Committee in Beijing.

Environment, economics, and technology also influenced village decisions. Suburban units are usually rich and possess strong brigade economies. With these economic resources, which include capital, equipment, and access to employment in commune and brigade-run enterprises, brigade leaders influenced team decisions in ways that their counterparts in poorer areas could not. Suburban units also have less land per person; for them, incentive systems that sped up work were generally superfluous. But in hilly, more isolated regions, some peasants welcomed the new independence that came with household liability for agricultural production (*baochan daohu*).

Cropping patterns directly affected the incentive systems that were chosen. Vegetable production, with its numerous varieties, creates complicated accounting and organizational problems when individual quotas are used. Wet rice cultivation, particularly in hilly areas, requires complicated irrigation schedules and careful coordination among households. In these situations rationality favored continued

Table 6.1

## Major Characteristics of Three Research Sites in Jiangsu Province, 1980

| Brigade | Distance from Nanjing (km.s) | No. of teams | No. of villages | Major crops | Per capita income (yuan) | Population | Acreage/ hectare | Land/labor ratio | Terrain | Level of mechanization: no. of tractors |
|---|---|---|---|---|---|---|---|---|---|---|
| Mushuyuan | 2 | 7 | 4 | vegetables | 320 | 1,594 | 35.5 | .06 | flat | 17 |
| Qingxiu | 30 | 26 | 14 | rice,wheat | 230 | 3,666 | 424.2 | .37 | mixed | 27 |
| Sept. 1st | 50 | 8 | 8 | rice, wheat, cotton | 120 | 1,824 | 114.3 | .21 | hilly | 9 |

collective incentives. But for cotton, where output depends on hand picking, and for other dry-field crops where no irrigation problems exist, household or individual quotas may succeed; the only exception occurs during droughts. Therefore, household contracts appeared first in areas more dependent on dry-field crops. The case studies demonstrate how variations among units caused leaders and peasants to respond to the new policy in quite different ways.

But local leaders' choices about incentive systems depended not only on local conditions, but on the special character of the leadership and economy of Jiangsu province as a whole. Jiangsu province probably sells more grain to the state than any other province in China.[5] Maintaining this high output needs continuous water projects. Also, the development of brigade- and commune-run enterprises has supported agriculture through the reinvestment of industrial profits. Therefore, high levels of capital formation, necessary to establish collectively owned industries and finance water and field improvement projects, have played a crucial role in the enrichment of Jiangsu province.

Many policies since the Third Plenum of December 1978 directly threatened the provincial leadership. Xu Jiadun, provincial party secretary until 1983, had built his reputation on Jiangsu's high grain output, both as the provincial officer responsible for agriculture in the early 1970s and as provincial party secretary. But restrictions on rural capital construction limited Xu from improving Jiangsu's irrigation system.[6] Distributing more money to the peasants[7] decreased the funds available for the expanding collectively owned enterprises and capital construction projects. And the emphasis on economic crops threatened Xu, who had been praised by Hua Guofeng and Li Xiannian for enthusiastically "taking grain as the key link."[8] Both of them, and particularly Hua, had helped to elevate Xu to the post of first party secretary after Peng Zhong had moved to Shanghai.

Using individual and household quotas as well as contracts exacerbated these problems by altering the relationship among peasants, local leaders, and state officials. Mobilizing peasants for capital construction projects through contracts would leave local units free to refuse to participate if projects were not to their benefit. Second, household or individual contracts would weaken the collective's ability to accumulate capital. And finally, surplus grain would fall under the control of teams and households threatening the province's continued high levels of grain sales.

Not surprisingly, Jiangsu lagged behind much of the country in establishing more individualized accountability systems. Only after

Zhao Ziyang or Hu Yaobang criticized him and told him to repent for his "leftist errors" did Xu Jiadun return to Jiangsu and convene a provincial meeting to discuss the new policies. He is reported to have made a public "self-criticism" there. From then until shortly before the Sixth Plenum in June 1981 when he publicly lauded the successful implementation of household production quotas in northern Jiangsu, Xu Jiadun's name rarely appeared in the provincial press. Lower level cadres, aware of the provincial opposition, probably felt little compulsion to change their incentive systems, and particularly to adopt more individualized ones.

## Mushuyuan: A Rich Vegetable Brigade

Mushuyuan Brigade sits right outside the western gate of Nanjing city. Comprised of four villages, it was Zijingshan Commune's model brigade, and the wealthiest brigade in the entire commune. For seven years it grew only vegetables and no grain. Every year the brigade lost rich cropland to an ever-expanding city, and the brigade party secretary believed that in approximately twenty years the brigade would disappear. I did my research there in January and February of 1981, though I occasionally returned until July 1981.

This brigade played a greater role in the daily activities of its teams and team members than most brigades in China. Although a 1978 experiment in brigade accounting was undone a year later, in 1981 all machinery was still owned by the brigade. Over one-third of the labor force worked in commune- and brigade-run enterprises. And for several years income from these industries had surpassed the total agricultural income of the teams, comprising 62.7 percent of total gross income in 1980 in the brigade. The brigade and commune had also underwritten a new sprinkler system, plastic tents, and other investments for agriculture, as well as a new old-age pension system.[9] Such extensive financial involvement probably created strong support for the brigade among its inhabitants.

This vegetable brigade was very slow to change its incentive system. The teams adopted "task rates" in 1978 but only for a few activities, and as of 1980 time rates were still used for most jobs. In 1980, the commune accountant advocated scoring workpoints after each day's labor ("flexible time rates," *sigong huoping*) to establish a closer link between quality of work and payment received. But team leaders, fearful that disagreements would arise daily, preferred to rank peasants once a month, if not once a year.

In fact, the vice brigade secretary believed that the leader of one team still allowed peasants in 1980 to rank themselves in open meetings (the Dazhai system of "self-assessment and public discussion") because he feared they would get angry if he set their workpoint values too low. When asked why he still used a system that the national government had rejected, he responded that changes were too troublesome. With so much work and so few laborers, time constraints made it simpler to let the peasants decide themselves. In fact, each field laborer in this team did work far more land than peasants in any other team, so the team leader probably was pressed for time.[10] But in 1981 the brigade pressured him to adopt time rates where the team leader set the workpoint values, which finally expunged the last remnants of the "leftist" Dazhai system in this brigade.

In January 1981, a brigade official told me that all teams were going to shift to task rates. In one team, I observed a freewheeling management committee meeting at which members discussed their personal preferences between time and task rates. Although four months earlier Central Document No. 75 had rejected time rates, males at this meeting argued vociferously for them. The women, on the other hand, strongly supported task rates. With time rates, number of days worked multiplied by relatively fixed workpoint values determined the number of points earned each year. But because workpoint values for women invariably were fixed below those for men, women preferred task rates, where the task's value was fixed. Though they might have worked slower, they would have had the opportunity to earn as much as men. Task rates, therefore, could have restricted sexual discrimination,[11] unless women were consistently given less valuable jobs. Although team management committee members freely expressed their views, they could not affect the final decision. Their team leader was extremely strong-willed and firmly held the reins of political control. During the Cultural Revolution he had led one faction that had usurped commune administration for a period of time. The day after the committee meeting he told me that he had decided unilaterally to use task rates, not out of fairness to women but because people would work faster. Nonetheless, all other teams still paid people on a time rate basis.

While their incentive system for organizing and paying general field labor failed to keep pace with national policy, this brigade was far more flexible in raising the vegetable seedlings that were subsequently transplanted into the fields. In 1978 it placed quotas on the agricultural inputs for the seedlings, and in 1979, soon after the Third Plenum sanctioned it, the brigade gave a bonus for meeting the quotas. The bonus, given to the one or two technical specialists each team had for raising vegetable sprouts, was based on the quality and quantity of sprouts raised. The brigade also had a "technical" team that produced sprouts not only for its own fields but for other teams as well. The agricultural prosperity of the brigade depended heavily on all these technical specialists, hence the readiness to give bonuses. As one brigade official remarked, "Things that have been good for the countryside, we begin to do early on."

The brigade wanted to make workers in all its enterprises more accountable for their work but it had troubles in the brigade-run piggery. The women laborers there told me that they already worked hard enough. Also, they feared accepting personal financial liability for each piglet that died. In their experience, one piglet died in almost every litter. Due to this opposition, the brigade, as of July 1981, had still not established a personal incentive system in the piggery.

Even as team leaders began contemplating a shift to task rates, the brigade leaders began devising a new scheme. They had only alluded to it in January, but by March their proposal had taken shape. Two or three work groups within each team would work separately, with the total number of workpoints and the amount of money that the group could use for planting, raising, and harvesting specific crops fixed. Because any financial savings would revert to them, the groups would work harder and complete the tasks sooner. But if they failed to meet the input quotas they would be fined. However, the brigade still would not give them a production quota. This method, called "long-term task rates" (*da duan bao gong*), first appeared in public in the *Nanjing Daily* in the late spring, early summer of 1981, when pressure on all units to link income to output was mounting. By adopting long-term task rates, local leaders could argue that they were improving labor accountability without instituting output quotas.

By July, however, brigade officials said that due to the strong propaganda campaign in the press and other undefined pressures advocating linking income to output, the brigade would establish individual liability (*lianchan daolao*) for all crops linking income directly to output. Even though an accounting nightmare could arise, particularly because teams in the brigade no longer had accountants, the necessity of this move had become clear to them after a May 20 article in the *People's Daily*.[12] One brigade official had already gone to Hangzhou and Shanghai to see how other prosperous brigades linked individual income and output.

Commune officials agreed that pressure was intensifying.

The upper levels are pushing this policy hard now but around here it doesn't sit well (*chi buxia*). You can't do water conservation projects. How will you use the sprinklers and the machinery? . . . They are really going back. After so many years of building the collective system they are taking it apart. There is strength in the collective system because the numbers are big. In Anhui the policy that they are using now is not good for us. We do not want it here. . . . If you start any type of system of responsibility and then it doesn't work, well, okay. But if the weather is bad one year, they may shift to "household quotas." We don't like that here, so we are a little afraid of it.

Due to high per capita income and a stable brigade economy, team and brigade leaders were comfortable using time rates. Nonetheless, they changed this local policy because of pressure from above. After Central Document No. 75 had been circulated in the winter of 1980–81, they began the transition to task rates. And although they were frightened of it, by July 1981 the brigade was preparing to cross the threshold and join other units across the country that were linking individual income to the quantity of crops produced.

## Qingxiu: A Rich Grain Brigade

Qingxiu Brigade is situated in the southwest corner of Jiangning Commune, Jiangning county, thirty miles south of Nanjing. On the east it borders the Jiangning River, which flows directly to the larger Yangze River. Due to massive water conservation projects in the 1960s, most teams had an abundant water supply. Grain production was high, surpassing 15,000 kgs per hectare. According to the brigade secretary it sold more grain per capita to the state than any other brigade in Jiangsu. Therefore, it benefited greatly from the 50 percent price increase in 1979 for above-quota grain sold to the state. In 1973, when Xu Jiadun heard about its extensive land leveling projects, he made the brigade a provincial model. Leadership had been relatively stable. After many years the old brigade secretary became an official in a neighboring commune, and the secretary in 1981 had managed the brigade firmly since 1975. I lived in this locality for fifteen days in June 1981.

As in the vegetable brigade, the role of this brigade in the economy of the teams was very large. Since 1978, the gross income of the brigade's five enterprises had equaled or surpassed the total agricultural income from the twenty-six teams. Including laborers in commune-run enterprises over which the brigade had veto power, over one-third of all laborers worked in collective industries. In 1981, factory workers in the brigade were still paid workpoints and their salaries were remitted to their respective teams for distribution among all members. These funds were crucial for maintaining the brigade's impressive per capita income of 230 yuan, the highest in the commune. [13]

All teams employed task rates (*bao gong*) for harvesting wheat and rice and only during the 1968 "Three Loyalties Campaign" did they cease using them. At that time, they also introduced the system of "self-assessment and public discussion." Although people worked harder in the first year, these initial successes resulted from the "milieu of activism" generated by the political campaign, rather than the particular incentive system. [14] According to one brigade official, they returned to time and task rates in 1971 because

> If you don't give people workpoints for a long time they don't work. In the end there was no benefit [from Dazhai workpoints], so we stopped in 1971. To start the policy the upper levels called a meeting. This was the commune. They had learned about it from the outside. When we stopped, we did it on our own. The whole brigade stopped. Production wasn't good, and the masses said they were working more but not getting more. They said that "Dazhai" workpoints had become "dagai" ["approximate"] workpoints.

In 1981, many production teams extended task rates to the work group within the team. Each group received workpoints, based on the area of land they worked that day, which were then redistributed to the individuals in the group based on time rates. A few units adjusted this procedure by daily evaluating each individual's work so that the quality of work directly affected income. The team leaders employed this method only for rapeseed because if not well planted, it will die.

The first link between income and output occurred for rapeseed cultivation. In fall 1980 the leader of the wealthiest team gave output quotas to two groups (*lianchan daozu*), with a 30 percent bonus-10 percent fine system. For two years many seedlings had died, but with this method the 1981 harvest improved significantly. He planned in 1982 to give two groups output quotas for wet wheat and rice fields as well. Unlike many teams in the brigade which are in the hills, this team had flat land. Irrigating each group's wet fields would not cause problems.

Although his team was the richest in the brigade, the team leader, rather than conservatively avoid risk, responded as a "rational peasant," [15] using his team's financial surplus to try out new ways to get even richer. For many years this team had been the test point for all new seeds in the brigade. Although this too involved risks, his team benefited first when the strains were successful. His experiment with "group quotas" had already triggered a "demonstration effect." A neighboring team leader, related through marriage to this risk taker, knew of his successful experiment in rapeseed and told me that he would give output quotas for rapeseed to each household (*baochan daohu*) in fall 1981.

Nonetheless, most team leaders opposed the policy and argued that their team members were unhappy with the prospect of using it. The possibility of a fine frightened them most. The peasants felt that on their own they would be unable to meet quotas based on the high per-hectare output achieved by the collective. The leader of another prosperous team expressed it this way:

> They're not willing. Now they have machines, so the work is lighter. . . . They don't mind more private plots, but if they "fix production" (*baochan*) then they have a target they must meet. I don't want to do it either. I called a team meeting and told the peasants that anyone who wanted to work on their own could do so. . . . But no one wanted to do it. Now the output per hectare is high and they would have to meet the output that the team gets.

The brigade's attitude toward this policy was unclear. One official told me that they supported the policy, but the way it was introduced into the brigade left some doubts as to their true feelings. In March 1981 the county announced that poorer hilly areas could institute household production quotas.[16] On returning from this meeting brigade officials reportedly confronted peasants on the street, asking them if they wanted to work the land on their own (*dangan*). One person said, "they blew a wind" (*gua yige feng*), implying that they triggered a mass response. In most cases it was fear. When the team leaders were called together and invited to try out the policy there were no takers. When next the poorest team in the hills was offered the chance to experiment with individual quotas for wet rice, they resisted. Brigade officials told me that only 10 percent of the people supported the policy.[17] Of the remaining 90 percent, 60 percent opposed it passively and 30 percent opposed it *totally*. "They curse it," they said. "These people either had many family members working in collective enterprises or were weak laborers."

Though Western analysts assume that all peasants prefer private farming, I found serious hostility toward any prospects of this transition. I first talked with a peasant in a team where, due to the increase in the number of peasants working in brigade factories, per capita income had not dropped from 1979 to 1980. His family had three field laborers and two factory workers, but he claimed to have an average income for the team. I asked him the one request he would have for an investigation group from the party center. His preferred financial situation notwithstanding, the frankness of his answer is rather surprising.

> Before we were so poor and now the economic work is done well. The new policy is a backward retreat. If you divide the land you will have "polarization" (*liang ji fen hua*). We will have the return of the landlords and capitalists. Collective socialism is good. Chairman Mao is good. We do not want to move off of Chairman Mao's road. The collective is done well. If you divide the land to the household, how will we use the machinery? If you divide the land to the household then the team leader will have no power and the place will be chaotic. Even the households with five laborers do not want to divide the land.

His attitude was corroborated by a peasant in another team. Due to the brigade's liberal policy of the last two years on opening barren land as private plots, he farmed as much land in 1981 as he had before land reform. His relatively stable income from the collective was supplemented by his private plot, so he opposed dividing the collective fields. Interestingly, he also extolled "Chairman Mao's socialist line." A peasant in yet a third team, who also had greatly expanded his private plot, preferred collective farming because it was easier simply to work where and when they told him to.[18]

One team leader attributed the peasants' enthusiasm for the collective to the brigade enterprises.

> The peasants here work hard. They do their work quickly. If I give them one day to do the work, they do it in half a day. So, if the people want to go work in their private plots, they can. But most prefer to work on the collective fields. This year our income from sidelines increased by 7,000 yuan, mostly due to the fact that we sent more people to the brigade enterprises. Last year we had twenty-one people in the factories and now we have sixteen more. *The collective income will increase and the peasants know this. They can calculate. They know the collective income will increase.*

The issue of mechanization, which surfaces in two of the above quotations, also explains the peasants' antipathy toward the new policies. In

1981, after this brigade became a test point for mechanized farming, the County Agricultural Mechanization Bureau donated ten 12-horsepower tractors. The brigade had already purchased fifteen tractors, so with this supplement all twenty-six teams had at least one 12-hp tractor. In this brigade, which was the only place I visited where ploughing was done primarily by tractor, the prospect of tilling land by hand was hardly an enticing thought.

Commune officials agreed with the peasants and the brigade officials.

> At present, output in our commune is very high for the county. Our technical level is also high. Our rice output is especially high. So, on this basis, we recently began to say that if the masses want production quotas to the household or the labor power it could affect per capita output. But the masses in the whole commune are not willing to have production quotas to the household or the laborer for wet fields. They do not have this desire. They still want to do it on the same basis and in that way increase production and mechanization.

Nonetheless, the commune permitted one brigade which grows cotton and wheat in dry fields to establish household quotas. The brigade had never changed these fields into paddy, and with the restrictions on large-scale field construction, output could be increased only by adopting the new incentive systems.[19]

As with most of Jiangsu province, Qingxiu Brigade was not at the forefront of organizational change. The hesitancy of the province was reinforced by resistance to household quotas from the commune and probably the brigade leadership as well. The high level of technology, the high grain output (which peasants felt they could not match privately), and collective enterprises that supported per capita income in the teams made the peasants fear changes in the well-established collective system. Also, larger private plots and increased sideline incomes made the incentives in the household quota system less enticing. Nonetheless, the most prosperous team was willing to risk experimenting with new incentive systems. And these changes occurred first for rapeseed, a dry-field crop involving careful planting but not complicated irrigation procedures.

### September First: An Average, Hilly Brigade

The September First Brigade lies in Tangquan Commune, Jiangpu county, on the north side of the Yangze River. Though still located in Nanjing municipality, it borders the Qu River, with Qu county in Anhui province on the other side. Though its per capita income of 120 yuan was 50 percent higher than the national average, the September First Brigade was the poorest brigade in the commune. Disparities within the brigade were also significant; two of its eight teams were short of arable land and received a state grain supplement. Peasants in these teams grew grain in their private plots to avoid purchasing it at state shops. Though the irrigation system was adequate, half the land in some teams could grow only dry-field crops.

Leadership in the brigade had not been outstanding. Brigade officials had lacked the prestige and authority of the leaders in the other two brigades. While I was there for a month from April to May 1981, one vice brigade secretary, who was illiterate, retired. His replacement, a young ex-army officer who had led one team for three years, was selected in order to add vitality to the leadership ranks.

Economically, the impact of the brigade on the production teams was far less significant than in the two brigades discussed previously. The only brigade-run enterprise, a pig-bristle factory, was donated by the commune in 1975 because only this brigade, out of thirteen in the commune, had no factory. The teams had purchased almost all their own machinery; the county had supplied two tractors to the poorest teams. And in 1976–77, when the factory paid the peasants workpoints, it had remitted less than half as much to their respective teams as in the model grain brigade. In 1981, due to low monthly wages, few peasants sought employment in the factory, preferring work in the fields or outside the collective. Only 13 percent of the brigade's total labor force worked in the factory, mostly young women, the handicapped, and the weaker males. This economic ineffectiveness led one team leader to refer to the brigade as "an empty shell."

Throughout much of the surrounding countryside the peasants responded quickly to the opportunities to farm on an individual or household quota basis. In spring 1978, as agricultural policy was just beginning to change but more than half a year before the Third Plenum, thirty-nine of the ninety-three teams in the commune established household production quotas for cotton. Although the county tried to prevent this, commune officials failed to tell the brigade leaders of the criticism they had received for permitting these changes. "If I had said anything to the brigades, they would have stopped. So, I simply kept quiet." And in 1979, when the county remained silent, the scope of household quotas increased.

But in early 1980 antagonism toward household quotas intensified in Jiangsu. The city and county governments intervened, demanding that local units stop using individual incentive systems for crop production that linked income to output. They accepted contracts only for sidelines (*zhuanye chengbao*). This time commune officials passed on the instructions, though still suggesting that units already having production quotas for groups at the subteam level continue to use them. Household quotas, however, had to be discontinued. But by fall 1980, almost all the commune's teams were using household quotas for cotton and rapeseed, both dry-field crops. And after the *People's Daily* editorial of March 2, 1981, which for the first time supported individual production quotas for wet fields in average income areas, six of the commune's poorest teams decided to experiment with this policy.[20] Forty-seven teams had agreed to shift to group quotas for wet fields, while the remaining forty decided to stick with task rates. As the policy options expanded, people and leaders in this locality willingly partook of the new innovations.

Peasants here began to use household quotas in 1978 because of this area's proximity to Anhui, which was one of the first provinces after Mao's death and the fall of the Gang of Four in 1976 to experiment with these new incentive systems. News from Anhui travels quickly because of livestock marketing across the provincial border and extensive personal ties.

The peasants and leaders in this hilly brigade, however, did not respond as quickly. In 1978, they still planted as a team using task rates, and in 1979 only four of eight teams had household quotas for cotton. Under pressure in 1980 the brigade retreated; three teams followed the city's guidelines, using special contracts for raising cotton, but the rest cultivated the cotton fields collectively.[21] According to commune officials, this brigade, as compared to others in the commune, panicked in 1980 and overreacted to the outside pressures. Only in 1981, after the political environment sanctioned it, did the brigade uniformly adopt household quotas for rapeseed and cotton.

Even weak leadership, which made the brigade vulnerable to external political pressures, fails to explain its slow response. With the lowest per capita income in the commune, the peasants here should have favored these incentive systems in 1978, particularly since agricultural output and per capita income had dropped in 1977. They knew of the successes across the provincial border. Moreover, unlike the two other brigades where the richness of the collective caused peasants to resist the new acountability systems, this hilly brigade's economy was less significant. Changes in the collective system should not have frightened the peasants. Nonetheless, technical problems and mistrust of the party's future policies were at work in this brigade. Also, some peasants were more interested in increasing sideline income than collective agricultural output.

Agricultural policy in this area has been volatile for twenty-five years. During the Great Leap Forward the commune collectivized and sold to the state private land that the peasants had planted with trees just a few years earlier. During the 1968 "Three Loyalties Campaign" peasants lost their private plots for three years. In 1975 and 1976, responding to political pressures from a "leftist" county party secretary, local officials uprooted crops in some private plots. And in 1980, commune leaders again tried to eradicate private tree-seedling cultivation. The swiftness with which the incentive systems were introduced in this locality probably made the peasants cautious. According to the leader of one team,

> The peasants still do not trust things. At the meetings they ask me how long the policies will last. One year? Two years? Five years? It's important for household quotas because if peasants trust you and the policy they'll exert more energy and will plant deeper. In this way, in dry seasons the roots will be deep and they'll get more water. Also, the wind won't blow them down. In the wet season, if the roots are deeper they won't get waterlogged.

The technical problem of irrigating uneven wet fields also played a role. In a December 1980 meeting the county announced Central Document No. 75 at a meeting. After brigade officials reported to their teams, one team prepared to establish group quotas for wet rice cultivation. But after deciding precisely on what type of system of payment to use, they hesitated. Their paddy fields, created by leveling hilly land, were uneven in height. With only one portable pump capable of raising water six feet, some fields got water later than others. The conflicts that arose during the planning sessions convinced the team leader that the problems exceeded the potential benefits. So, they planted the rice as a village with task rates. Another team also thought of using group quotas for their wet fields in 1981, but irrigation problems led them to defer as well. With the poorest water supply in the brigade, a shortage of water would have caused the groups to fight.[22]

Another technical problem was the inexperience of some young peasants. After raising rice seedlings in densely packed clusters, great care is necessary when digging them up, separating them, and replanting them in neat rows within the paddy fields. Lacking these skills, young people would not have gotten a good harvest and might have faced a fine. For this reason, the brigade opposed household quotas for wet fields.

Other examples of hesitancy and risk avoidance occurred in the brigade. In 1979 the wealthiest team rejected household quotas and used only group quotas for cotton because with task rates its 1978 cotton production had surpassed all teams in the brigade. At the other extreme, the poorest team in the brigade had grown accustomed to being poor and receiving state assistance. After losing ten hectares of land during the Great Leap Forward for the building of the commune reservoir, it could not grow enough to eat. Since 1963 it had relied on state grain subsidies. In line with 1981 policy, whereby peasants in teams relying on the "three supports"[23] were permitted to farm individually without any quotas (da baogan), the commune offered this opportunity to this poor team.[24] But, after discussions with the peasants, in spring 1981 the team leader decided not to do it.

> Most people oppose the policy because there would be no collective economy. People fear "total responsibility" (da baogan). They like the collective method [household quotas] because they get fertilizer, insecticide, and thirty-five points for managing the fields, all from the team. If they get sick and can't plant, the team plants for them. Also, they will still get their grain ration. They will lose their workpoints but they can borrow money to buy their share of "collective grain" (gong liang), and pay back the money when they can.

With "total responsibility" they would probably produce enough grain to feed themselves, but it would obviate state support. The insecurity of freedom can be a frightening experience, especially when one is accustomed to being coddled by the state. As a result, they rejected all policies linking income to output until 1981 and then only for cotton and rapeseed.

Private sidelines might have affected the adoption of household quotas because after 1977 the number of peasant families raising tree seedlings increased greatly. Peasants drawing out their personal capital to invest in tree-seedling cultivation might hesitate to accept household quotas due to the potential fines. But in spring 1979, when four teams decided to plant cotton on a household quota basis, the number of households raising tree seedlings in these teams also increased, often significantly. Perhaps the situation was reversed: peasants planning to invest in tree seedlings hoped for a bonus to help pay off their new investments. And due to low collective agricultural output, peasants here may not have feared the fine.

Land/labor ratios could play a role in the decision on incentive systems, but the fit is not good either (table 6.2). Undoubtedly, Team 3 needed a remuneration system that quickened the pace of work, but in 1980 it reverted to collective labor and not group quotas. Also, based on these ratios, why did Team 6 adopt household quotas and Team 1 did not?

Table 6.2

**Land/Labor Ratios, September First Brigade, 1979**

| Team No. | 1 | 2 | 3 | 4 | 5 | 6 | 7 | 8 |
|---|---|---|---|---|---|---|---|---|
| Land/labor ratio (in hectares) | .25 | .23 | .31 | .24 | .25 | .11 | .12 | .20 |
| Adopted household quotas for cotton | No | No | Yes | Yes | Yes | Yes | No | No |

The question of leadership explains some of these anomalies. The strongest leaders were in teams 2, 4, 5, and 6. Lineage conflicts made Team 8 impossible to run, while in Team 7 the old leader stayed in office only because no one else was willing to manage this poor village. Similarly, leadership had been weak in Team 1. Many people told me that the team leader, often criticized as a "rightist," feared the political bureaucracy. He was precisely the type of leader least likely to experiment with a new incentive system. Lastly, Team 3 did not adopt any quota system in 1980 due to a leadership change. Therefore, only teams with strong leadership were willing or able to adopt household quotas.

Also, the two teams most dependent on sideline occupations for their collective income, 7 and 8, were the slowest to adopt any new accountability system (table 6.3). Both had links with outside units, such as construction and forestry companies, and sent excess laborers to work there. The change since the 1978 Third Plenum that interested them most was undoubtedly the new liberal policy on sidelines. Only after Central Document No. 75 made the incentive systems official policy did these two teams finally adopt household quotas for rapeseed and cotton.

Compared to the other two brigades studied, the great variations among the teams in this hilly brigade and the yearly shifts within teams

Table 6.3

**Ratio of Total Collective Profits Derived from Sidelines**

|  | 1973 | 1974 | 1975 | 1976 | 1977 | 1978 | 1979 | 1980 | Average |
|---|---|---|---|---|---|---|---|---|---|
| Team 7 | .62 | .53 | .67 | .69 | .54 | .61 | .62 | .60 | .61 |
| Team 8 | .40 | .43 | .58 | .50 | .62 | .83 | .62 | .75 | .59 |
| Average of other six teams | .36 | .33 | .35 | .23 | .35 | .37 | .37 | .29 | .33 |

are most striking. The brigade leadership reacted swiftly to external intervention and, in the absence of those pressures, each unit's or leader's characteristics determined the response to the incentive system. And with large interteam variations, local policy within the teams was rather different.

Certain trends, however, continue. Major innovations occurred first with dry-field crops. Technical complications, inherent in irrigating paddy fields in hilly areas, again caused hesitation in establishing household or group quotas. And once more, peasants were nervous about the new changes. Interestingly enough, however, the feebleness of the collective did not increase peasant support for its further weakening. For peasants in the September First Brigade the subsistence guaranteed by the state and the collective provided a welcome base from which they could increase their household sideline income.

## Conclusion

Though in many localities the new incentives are quite popular, not all peasants warmly welcomed these adjustments to the collective system. Opposition lay in wealthier, more developed units, where peasants knew that their household's prosperity was intertwined with a strong collective. Even as national leaders downplayed mechanization, peasants in areas where machine use was practical still believed that this was one way they could escape from the drudgery of agricultural labor. Ironically, these were the units that in the 1960s and '70s learned well from Dazhai: they had built extensive irrigation systems, had leveled fields, and had sacrificed immediate aggrandizement in order to speed up capital formation and build strong collectives. In 1981, peasants in these areas were hostile to policies that threatened those hard-earned gains.

Cropping patterns and terrain consistently affected policy implementation. In both grain-producing brigades the first breakthrough in household quotas occurred in the dry fields, and even after leaders overcame political fears, problems with irrigating paddy fields still caused teams to hesitate. Also, because vegetable-producing units cultivate a wide variety of crops, leaders there eschewed individual quotas as long as possible. Psychological factors too were not unimportant, but peasants and local leaders mistrusted the permanency of the changes and feared future policy shifts much more in the hilly brigade, where policy has been more unstable and unpredictable.[25]

Twice the bureaucracy intervened in the decisions of these rural units. Before the publication of Central Document No. 75 in September 1980, provincial, municipal, and county officials sought to prevent experimentation with incentive systems linking income to output, even if it could help agricultural production. Units with weak leadership succumbed swiftly to these bureaucratic interventions. But for wealthy units who themselves opposed the changes, the middle-level hostility to the policy increased their resolve to persist in their own opposition.

After September 1980, however, the pressures began to change. As direction from Beijing intensified and became more specific, some units that preferred to move slowly found themselves under pressure to introduce quickly new incentive systems. Seeing that the tide had changed, middle-level bureaucrats swung full circle and began demanding policy conformity, once again oblivious to the issue of local independence and economic rationality.

Finally, major problems result from the concomitant implementation of broad policy changes that criss-cross a large spectrum of issue areas. The unintended consequences of one policy may undermine the goals of another. This fact may explain why policy shifts in China, which often have been so all-encompassing, quickly run into unexpected complications. The leaders in Beijing may have correctly assumed that peasants would respond to these new incentives. But liberalization in other spheres, such as the expansion of private plots, resurrection of rural trade fairs, and increased freedom to pursue household and collective sidelines, may have made this new policy less attractive. The incentives from the former sectors involved fewer risks and greater opportunities for increased household incomes than the inducements built into these accountability systems, particularly those that involved the meeting of a production quota. Therefore, more rapid and voluntary implementation of household, group, and individual quotas should have occurred in areas with few avenues for increasing private or collective sidelines. But, when peasants were able to get rich by methods other than laboring in collective fields, we should not be surprised to find that they did not warmly welcome these new incentive systems.

## Notes

1. See David Zweig, "Opposition to Change in Rural China: The System of Responsibility and People's Communes," *Asian Survey* 23 (July 1983):879–900. For the development of new agricultural policies at the national level, see Jurgen Domes, "New Policies in the Communes: Notes on Rural Societal Structures in China, 1976–81," *Journal of Asian Studies* 51, 2 (February 1982):253–67, and David Zweig, "Content and Context in Policy Implementation: Household Contracts in China, 1977–1983," in *Policy Implementation in the Post-Mao Era,* ed. David M. Lampton (forthcoming).

2. The term "local leaders" refers to team and brigade officials who differ from "middle-level cadres" at the commune, county, district, and provincial level because they are not state-salaried officials. Their income depends directly upon the production levels in the village or villages they manage.

3. This phenomenon, called *yi dao qie* or "one cut of the knife," has been criticized since the Third Plenum of December 1978. During the "Movement to Learn from Dazhai in Agriculture" many localities, whose natural conditions differed greatly from Dazhai's, still had to do things the way Dazhai did.

4. See the chapters by Victor Nee and Jonathan Unger in this volume.

5. According to an assumed model of the interregional movement of grain, Jiangsu has the largest surplus, with Heilungjiang next. See JETRO, *China Newsletter* 34 (September-October 1981):16.

6. During a visit to the Ministry of Water Conservation in Beijing in early 1981, I was told that half the staff was busy deciding which projects to cancel. Jiangsu took a public position against cutbacks on water conservation projects in *Red Flag.* In this article the authors attributed Jiangsu's prosperity to continued improvements in the irrigation system and stressed that, unlike other provinces, Jiangsu's large projects had been well managed and not wasteful. See "An Investigation in Jiangsu," *Red Flag,* April 1980.

7. The movement to "lighten the burden of the peasants," part of the Xiangxiang Experience propagated by a central document in summer 1978, signaled this change. See *People's Daily,* July 5, 1978.

8. Much of this information on Xu Jiadun was obtained from personal conversations with well-connected Chinese in Nanjing.

9. Begun in 1981, this system allowed women to retire at age fifty-five and men at sixty and gave them .80 yuan per month for each year worked. Thus a peasant who worked for twenty years received 16 yuan per month.

10. For 1980 the average land/labor ratio for the brigade was .06 hectares per person, and the standard deviation of all the teams was .017. Therefore, the land/labor ratio in this team of .09, the highest in the brigade and almost two standard deviations from the brigade mean, indicates that the difference between this team and the others was rather considerable.

11. A commune official told me that although this was official commune policy, as of January 1981 only 20 percent of the teams were giving women equal pay for equal work, implying that the vast majority of teams in the commune still used time rates.

12. See "Time To Consider 800 Million Peasants," *People's Daily,* May 20, 1981, p. 1.

13. From 1979 to 1980 average agricultural income of all teams dropped 33

percent, but per capita income decreased by only 4.8 percent. This occurred because in the same period the average increase in team industrial income, which was derived almost totally from these remitted salaries, was 37.5 percent. To demonstrate the importance of these factory jobs to the high standard of living in Qingxiu Brigade, I regressed the percentage change in industrial income, the percentage change in agricultural income, and the percentage change in population, all for 1979–80, on the percentage change in per capita income from 1979 to 1980. I did this for all twenty-six production teams in the brigade. These three independent variables explain a great deal of the variation in the change in per capita income ($R^2 = .594$). Also, not surprisingly, the influence of agriculture on per capita income remains most important, with a standardized least squares coefficient of .725. Nevertheless, industrial income also had a great effect on the per capita income in each team. Its SLS coefficient was .560. Without these factory jobs, all of the peasants would have been significantly poorer. While results were significant at the .01 level for both these variables, it was not significant for population change.

14. See Jonathan Unger, "Collective Incentives in the Chinese Countryside: Lessons from Chen Village," *World Development* 6 (1978):583–601.

15. See Samuel L. Popkin, *The Rational Peasant* (Berkeley: University of Califronia Press, 1979).

16. This position meshes more with the political line advocated in the March 2 editorial in the *People's Daily* than with Central Document No. 75 of fall 1980, because the poor, hilly areas in Jiangning county were still twice as prosperous as the national average. They did not fit the criterion of "poor" units as outlined in Central Document No. 75.

17. When I accused them of triggering opposition by offering individual rather than group quotas, which given the financial level of the team seemed inappropriate, they explained that this team had a history of factionalism. The work groups would have divided on factional lines, creating endless, bitter conflicts over water and machinery.

18. The party secretary of a brigade at my third research site liked the new incentive systems because they helped overcome precisely this attitude. I translated a little ditty he recited for me that reflects this problem: "When the bell rings, begin your task./ Questions on the job? The team leader you can ask./ After work, go home/ and just relax."

19. They had planned to transform the sandy soil into paddy to increase the commune's total grain production. But in April 1979, when the Third Plenum documents were presented for discussion, the county stressed that "natural conditions must be respected" (*yin di zhi yi*). The commune cancelled its plans for changing the fields. Then, in the fall of 1980 after the September drafting of Central Document No. 75, the commune decided to let the brigade establish household quotas.

20. See "Summarize Experiences, Improve and Stabilize the Agricultural System of Responsibility," *People's Daily*, March 2, 1981.

21. Limiting contracts to sidelines did not really end the use of household contracts for cotton. In 1980, under specialized contracts, individuals accepted responsibility for meeting quotas and received bonuses for surpassing those quotas. In "form" they stopped household quotas, but in "reality," by using specialized contracts, individuals within households still contracted to produce cotton under an agreement that linked their remuneration to output. Policy evasion at the local level in China is often based on the difference between formalistically implementing a policy and really doing it. See the concluding chapter in my doctoral dissertation, "Agrarian Radicalism in China, 1968–1978: The Search for a Social Base," University of Michigan, 1983.

22. Liuhe county in northern Nanjing used group and household quotas for wet fields in spring 1981. The northern part of this county is reported to be quite poor. In April and May the area suffered a drought, and according to the *Nanjing Daily*, peasants, work groups, and collectives all battled over water.

23. The "three supports" refers to reliance on state grain for eating, state loans for production, and state aid for living expenses.

24. Even though there are no quotas, peasants are still accountable for meeting the state agricultural tax and for making small payments to the collective welfare fund. In most cases, however, the state tax was already waived in these poorer localities. For an analysis of these contracts, see Frederick W. Crook, "The 'Baogan Dashu' Incentive System: Translation and Analysis of a Model Contract," *China Quarterly* (forthcoming).

25. One peasant, who with his brother and brother-in-law made over 5,000 yuan raising tree seedlings in 1980, admitted that he would not be surprised if he were again criticized. He had simply decided to make as much money as quickly as possible. For this reason he had stopped working on the collective. When Zhang Jingfu, first party secretary of Anhui, discovered during a rural investigation that Anhui peasants still feared future changes, he announced that the Party Committee had no thoughts of making peasants change. "The length of time we had set will not change. Before we had set a three-year limit. But, if the peasants do not want to change, we will not ask them to change. Only if the peasants ask for change can it then be changed." See *People's Daily*, June 9, 1981. By 1984, Central Document No. 1 on agriculture announced that contracted land would not change for fifteen years. See ibid., June 10, 1984.

# 7

# Peasant Household Individualism
# Victor Nee

My interest in peasant household individualism was stimulated by the crisis in collective farming that took place in a village where I conducted field work in the spring of 1980. During the course of this field work, I was impressed by the basic success of collectivization in improving the condition of life for villagers in Yangbei. The establishment of the cooperative health plan and medical clinic brought effective medical care to the village, cutting back drastically on infectious disease and infantile mortality. The education system, substantially expanded during the Cultural Revolution, brought public education to the village, increasing educational attainment among younger peasants. The local government and village organizations appeared committed to local development, and they were generally staffed by cadres who impressed me as well meaning and dedicated. Finally, the improvement in the quantity and quality of factors of production developed after collectivization was substantial. Construction and maintenance of roads, development of new seed strains, availability of low-interest loans to purchase tractors, construction of a small hydroelectric plant that brought low-cost electric power to the village, the county-run agrotechnical station, the meteorological station that provided accurate weather forecasts, and many more smaller improvements have been part of a broadly based, state initiated modernization program in Wuping county. These improvements benefited peasants in Yangbei.

The author wishes to express appreciation for the careful reading and criticism he received from Randolph Barker, George C. Homans, Alan Richards, Steven Sangren, Mark Selden, Thomas Wilson, and especially William Parish. The field research for this study was funded by an international postdoctoral grant from the Social Science Research Council, 1979 to 1980. A faculty research grant from the University of California, Santa Barbara, provided funds for initial analysis of the field data. The Rural Development Committee and Center for International Studies at Cornell University provided a congenial atmosphere for work.

Despite these gains, during the course of my field work both peasants and cadres told me that certain farmers in the village still preferred to farm as single households and were dissatisfied with collective farming. During my stay in the village, these farmers were not a vocal group, constituting a seemingly silent opposition to the efforts to make collective farming more efficient and productive. Instead I sought to understand the reforms underway in the village at the time of my arrival. To be sure, there was considerable debate and discontent among villagers during my stay. This surfaced during the course of team meetings and in interviews with villagers. But it appeared to be articulated in the context of implementing reforms to solve the problems villagers complained about. With the reduction of the size of production teams, the policy to decentralize control to teams, and introduction of more effective incentive systems, it seemed prudent to assume that the basic features of collective farming would continue in Yangbei as they had since the mid-1950s. It was not until my return to the United States that I received a stronger confirmation of a more widespread preference for household production among peasants in Yangbei. A team cadre sent me a letter that informed me of developments in the village since my departure. I learned that the reduction in the size of teams had proceeded rapidly, resulting in one team left with only two households. This was in sharp contrast to the size of the old teams, which ranged from twenty to thirty households. From another letter, which I received in the spring of 1981, I learned that teams had divided up their assets and assigned land to individual households.

Before I left the village, the villager who later wrote to me told me that there would be further reductions in the size of teams than what I had witnessed during my stay. I assumed that liberalization of state controls would result in a situation similar to the period after the Great Leap Forward, when China adopted agrarian policies similar to the Soviet New Economic Policy. But I had not anticipated that decollectivization was on the agenda of reform. With the advantage of hindsight, I could better interpret the evidence in my data, especially those from my informant interviews, pointing to a persistence of peasant household individualism.

This return to household production in many areas of China necessitates reexamination of the peasant household as an economic unit. Since the mid-1950s, China had emphasized the development and consolidation of rural institutions as the means for accomplishing the modernization of agriculture. Analyses of rural development have for this reason focused largely on the four-tiered local organization system, composed of the county government, people's commune, production brigade, and production team.[1] By contrast, relatively little has been written on the Chinese peasant household as an economic unit.[2] Scholars assumed for the most part that the team, brigade, and commune system was a permanent feature of the rural landscape, and that the long-term trend pointed to further consolidation of collective modes of agricultural production.[3] Moreover, it was widely assumed that while the peasant household continued to be an important social unit for family life, marriage, and reproduction, as a production unit it played only a small role in the residual private sector, having been eclipsed by team and brigade management of agricultural production.

This chapter attempts to explain peasant preference for individual household production over collective farming. Although collective farming may still be in some areas the dominant mode of agriculture production, if the argument about the persistence of peasant household individualism is valid, I would anticipate a consolidation of the trend toward household cultivation, provided that the state persists in its current policy of relaxation of controls. The question I seek to answer is why, after more than two decades of state-initiated efforts to develop and consolidate collective farming, have peasants in many areas moved quickly to dismantle collective farming, choosing instead to farm as individual households?

This is not to say, however, that in all areas peasants would choose household production over collective farming, if the state were to relax all controls over agricultural production. The explanation of peasant preference for household production will also specify the likely conditions under which peasants may strongly prefer cooperation rather than going it alone. In these circumstances, cooperation is more likely to emerge from the productive process rather than as a result of state controls over agricultural production. Moreover, in these areas local officials in county and commune governments may actually have pressured peasants to adopt the household contract system so that the locality they administer remains in step with other localities where decollectivization has progressed rapidly.

There is little in the theoretical literature on peasants that provides a ready answer to why peasants prefer household production and under what conditions they will seek cooperative work arrangements. According to the moral economy approach, peasants would be expected to prefer the security associated with collective farming over the greater risks of farming alone, especially in poorer areas where peasants live close to subsistence margins.[4] If in China collectivization has failed to result in high growth rates comparable to those achieved in Taiwan, South Korea, and Japan during the same period, at least it did guarantee basic securities that Chinese peasants had not enjoyed in the past. If peasants are risk averse and seek to maximize security, why have the poorer mountainous and backward areas been at the forefront of the trend to dismantle collective farms? Samuel Popkin's "rational peasant" approach may be more useful in explaining peasant household individualism insofar as it assumes that peasants are rational in responding to incentive structures, understand investment logic, and are not necessarily risk averse in the pursuit of utility maximization.[5] Like Popkin, I draw on the insights provided by economists to explain Chinese peasants' preference for household production. I take an "economic approach" in the sense that I assume peasants, like other people, seek to maximize value through rational calculation of how to gain the optimum rate of returns for the resources they have in hand.[6]

Throughout the world peasant small farms have routinely achieved very high levels of efficiency.[7] The view that small farms in China also operated at high levels of efficiency has been argued by Chinn, and by Dittrich and Myers for the prewar period.[8] Berry and Cline observed that Taiwan farms of less than .5 hectares produce twice as much per hectare as do farms of 2 hectares.[9] If the present leaders in China and local cadres are concerned with raising agricultural productivity, small household farms are probably superior. I will argue in this chapter that it is the persistent belief that households can do better on their own that leads peasants to rush back to household production when given the opportunity by the state.

Insofar as the study attempts to explain peasant household individualism it seeks to develop a theory. As George Homans has maintained, "a theory of a phenomenon is an explanation of the phenomenon, and nothing that is not an explanation is worthy of the name of theory."[10] The first hypothesis follows directly from the above discussion.

Hypothesis 1: Peasants are more likely to prefer household goals than individualistic and community goals.

If hypothesis 1 is supported, then the importance of peasant household maximizing behavior is a matter of course. In hypothesis 2, I specify the conditions under which peasants will prefer household production.

Hypothesis 2: If the household division of labor is adequate as a production unit, peasants are likely to prefer individual household production over collective forms of production.

Implicit in hypothesis 2 is a specification of the condition under which peasants are likely to prefer cooperation. When the household division of labor is no longer adequate as a production unit, peasants are likely to prefer cooperative work arrangements.

Research for this village study was conducted in two stages. The first stage involved intensive interviews with former educated youths who lived in the village from 1969 to 1975. These interviews were conducted primarily at Cornell University, where I interviewed two young Chinese from Xiamen from December 1977 to September 1978. The interviews produced over 1,200 pages of interview text, which presented a surprisingly rich and detailed participant observation account of the years these former educated youths lived and worked in Yangbei village. Although the interview material was internally consistent and very credible, since my informants were not constrained in the candidness with which they retold their observation of life in the village, there was no way for me to confirm the objectivity and accuracy of their account without going to the village myself. I was able to do so in the spring of 1980, when I traveled to Wuping county in Fujian to conduct field work in Yangbei village. Accompanied by a former educated youth from Xiamen, who served as my research assistant and interpreter from Hakka to standard Chinese, I lived in Wuping county for one month. During this period I spent three weeks in the village, both to obtain an independent check on the data I had collected at Cornell and to collect additional data on subsequent developments in the village following the departure of the educated youths.

During the course of my field work, I conducted household surveys in four production teams, interviewed in my room peasants, cadres, and technical personnel, and had free access to all brigade and team statistical records. At the conclusion of my brief field work, I felt satisfied that my informants in Ithaca had given me an accurate and insightful participant observation account, and I was able to fill in gaps of data such as statistical information on population trends and economic performance and on developments since the departure of the educated youths from the village. To my knowledge, this was the first time that an American social scientist was able to conduct field work in a setting about which detailed data was provided first by refugee reports outside of China.

**The Setting**

Yangbei village is located in the southwestern corner of Fujian province in a Hakka district near the Guangdong provincial boundary. In a mountainous area, the village is considered quite remote. Despite recent improvements in roadways, travel to Yangbei from Xiamen still requires nearly two days. Within Wuping county, communications are surprisingly well developed, with daily bus schedules connecting Xiangdong commune with Wuping county seat and to the border towns in Guangdong province. Although provincial authorities consider the area backward, Wuping county is a rice-surplus area that sells grain to the state. Yangbei's standard of living is only slightly below the national norm, with peasant per capita grain consumption at 523 jin of unhusked rice per year, and peasant per capita income from the collective sector about 68 yuan in 1980. Yangbei in 1980 was a single-surname village with a population slightly over 2,400.

A case might be made that Yangbei's conditions are not dissimilar to those of other villages located in peripheral areas that are distant from central places. Certainly in a country where mountainous areas predominate, the number of villages in peripheral and less-developed areas is not inconsiderable. Moreover, Yangbei's political and economic integration following collectivization has followed essentially the same pattern as other villages in China. I do not, however, rest my case on whether or not Yangbei is representative. Not all areas in China have experienced the collapse of collective farming to the extent Yangbei has in the past years.[11] By explaining why collective farming came undone in Yangbei, I suggest that the propositions developed in this case study can be useful in explaining the return to household production in other areas of China.

## Continuing Importance of the Peasant Family

The collective economy in Yangbei guaranteed each household a basic grain ration, modest welfare funds for the poorest villagers, and inexpensive health care. However, as William Parish has argued, the peasant household still must fend for itself.[12] In the end the insurance provided by the collective was not that large, and the family integrity was still maintained as in the pre-Liberation village. This can be seen in the problems of the needy villagers who depend on collective insurance. Probably the most destitute members of the village are elderly men and women who do not have families to support and take care of them. The team supports these "five-guarantee" households. But at levels terribly close to subsistence, providing them with only their grain ration and no cash income. The "five-guarantee" households are expected to fend for themselves in other areas, such as tending their own private plot, raising poultry, and gathering twigs and straws for cooking. Though they receive medical attention from the brigade barefoot doctors without paying the annual fee to participate in the cooperative medical program, the team provides them with no other services. Neighbors may provide some temporary help in tending the vegetable plot or gathering fuel. But the single elderly have no assurance that help will be forthcoming when it is needed, or if it is, that it will be anything other than temporary, short-term, limited assistance.

On the other end of the age spectrum were a number of orphans whose parents died of malnutrition during the famine in 1960, and who have subsequently grown up. Very young orphans were adopted by kinsmen. Older children who were able to take care of themselves, however, continued to live in their parents' home. Their experience of growing up in the village mirrored the single elderly in the sense that they received very little informal assistance from neighboring households.

Divorced peasants try to remarry as quickly as possible due to the difficulties of living alone in the village. Women tend to remarry soon after divorce since by village custom, the house and custody of the children are kept by the man. Men also seek to find a new spouse soon after divorce. Household chores such as washing clothes, tending the private plot, and collecting fuel are considered women's work. Men who attempt to do women's work are subject to derisive ridicule by fellow villagers. Widows and widowers with older children are better off because the household tasks performed by a departed spouse can be assumed by their children. Nonetheless, they too seek to remarry. Widows with children, however, are less likely to do so since the marriage of a son would fill out the household structure, leaving the widow with security in old age.

The minimal welfare system combined with a lack of a tradition of charity, even between kinsmen, render peasants all the more dependent upon their family as a source of basic security and well-being. Sociologists have noted the weakness of the village community in presocialist China.[13] This feature of village society appears to have persisted in Yangbei even after collectivization. There was in Yangbei little evidence of a spill-over effect from cooperation in the collective economy into forms of informal mutual assistance between households. Peasants gossiped about hardships confronted by more unfortunate neighbors, but they did not try to provide help due to concerns that this might result in recurrent claims made on their own meager resources. Even prosperous households were vigilant in the conservation of household resources and avoided claims by needy kinsmen and neighbors.

The existence in the village of the very poor, comprising a stratum of single elderly, orphans, men without wives, and other irregular household arrangements, served to remind peasants of the importance not only of family, but also of having an optimal household structure. In Yangbei the typical households tended to be either stem or nuclear families. Households with joint families were uncommon, and they tended to be unstable. Household divisions generally occurred earlier than in peasant households in Taiwan or presocialist China.[14] The stem family provided a more favorable division of labor. In the stem family, grandparents helped out in taking care of the young, tending the private plot, performing household chores, and feeding the pig. This freed the mother to work in the collective fields to bring in additional workpoints for the family or to spend more time scouring the nearby hills for twigs and grass to burn in the family hearth. By contrast, the nuclear family experienced greater difficulty in achieving the optimal division of labor and tended to have a low laborer/dependent ratio. This was the case especially when the children were very young and the mother remained at home to take care of them and perform the household tasks, leaving only the father earning workpoints for the household. But the difficulties faced by the nuclear family were short term. Once the children grew up and began to contribute to the household economy, the nuclear family could also prosper.

To sum up, despite the guarantees provided by the collective economy, peasants continued to rely upon family-based strategies and calculations to achieve well-being and security. The fact that the collective economy was unable to provide more than the minimal subsistence-level guarantees, the difficulties of living alone, and the weakness of the peasant community were important bases for the persistence of peasant household individualism. Thus as long as villages supported living standards close to subsistence margins and collective welfare programs remained minimal, peasants continued to rely upon their households for their basic security and well-being.

## Household Utility Maximization Versus Community Goals

The moral economy school predicts that, when peasants are at the subsistence margin, the community will form a contract to provide a secure floor for all villagers. But George Foster, in his work on peasants and the limited good, suggests that the more common reaction among peasants is to view the world as a limited good or a zero-sum game.[15] If anyone else in the community benefits then it must be that I am losing. The degree to which this will be the perception varies around the world, but in China since collectivization in the mid-1950s there were many reasons for the zero-sum image to be an accurate one. With only very limited chances of moving out of the village or of starting nonagricultural enterprises in the village, everyone has been drawing on a single set of resources. The results are seen in how calculating peasants are in their relations with neighbors.

Peasant rational calculation tended to focus on maximizing individual household advantage over the interests of the collective economy. This manifested itself in a persistent problem, according to Yangbei cadres, in the complaint that villagers lacked genuine enthusiasm when working on the collective fields, by contrast to the effort displayed in the course of work on household private plots, sidelines, and household chores. This disparity between productivity in the collective and in private sectors points to the heart of the problem of collective farming in Yangbei. Simply stated, if all households benefited from the team economy performing well, then those who worked harder worried that their additional effort, though ultimately benefiting their own household, also might be subsidizing those who worked less hard. Accordingly, peasants put their best effort into their private plots and household sidelines, and were less likely to work with the same productive zeal and efficiency in the collective sector. This is the classic "free rider" dilemma.[16]

In fact, peasants are surprisingly fine-tuned in their calculations of household interests, something which the more egalitarian Dazhai workpoint system did not take into account. Moreover, they are highly responsive to incentive structures and demonstrate a subtle grasp of investment logic. This could be seen in the investment decisions centered on the purchase and sale of the household pig. According to Yangbei peasants, the timing of a man's purchase and sale of the family pig could result in greater or lesser profits given equal skills in raising a pig to full maturity. With the interest of seasoned investors, peasants followed the fluctuations of prices in the local markets to time their market transactions. On the basis of past experience, the peasant investor tried to anticipate periods when fresh pork was likely to be in short supply, and therefore prices higher. There are, however, many factors that influence price levels for pork. For example, a bad harvest year tends to result in tighter pork supplies and higher prices, whereas good years result in lower prices for pork. Thus a peasant must take into account the effect of weather conditions on the next season's harvest.

To complicate matters, the peasant investor must also try to anticipate seasonal fluctuations in the price of grain in the free market. As a rule of thumb, grain prices go up during the "spring hunger" months when peasants run short of grain allotted to them by the team. Initially, the price of fresh pork drops somewhat as supply increases when peasant households are unable to feed fully mature pigs and dump them on the local markets. But after this initial selling, the price of pork may climb when supply of pigs ready for slaughter declines toward the end of the "spring hunger" period. Thus the seasoned peasant investor in Yangbei sought to purchase the baby pig at a good price, buy and store free-market grain when grain was plentiful, even when the household had plenty in store for its immediate consumption needs, have sufficient grain supplies to last through the initial period of "spring hunger" to provide for both human and animal consumption, and finally sold the family pig as it entered into its high feeding period when prices were at their seasonal high. If prices for fresh pork in the local markets were higher than in past years, peasants responded to higher prices by investing in purchasing more grain and raising more pigs, and vice versa. However, not all households were able to take advantage of favorable price trends. A poorer household, for example, might not have adequate cash reserves to be in a position to invest in grain when it was relatively cheap. Instead it might be compelled to enter the market to buy grain when grain was most expensive. According to the supply-demand principle, grain prices were likely to be the highest when most households ran short of grain and needed to purchase additional supplies on the free market.

Peasants' marketing activity provides clear evidence of maximizing behavior, calculated to take full advantage of marginal profits through participation in external markets. Peasants in Yangbei are extremely attentive to price fluctuations not only in their own local market, but in the surrounding markets as well, including those in the border market towns in Guangdong. Village gossip is to a large degree centered on current market news. In the evenings men like to gather in a friend's house to pick up the most recent news about prices in the neighboring market towns. Women, as well, participate in the talk on market news, passing current news to their husbands, or using it for their own market activity. Peasant women often walk to Guangdong to sell vegetables and to buy grain or poultry for their own households to take advantage of more favorable prices in the Guangdong markets. Or they purchase items for their neighbors, receiving a small carrying charge for making the purchase. Men also walk to Guangdong to sell products from their household sidelines for slightly higher prices. The underlying logic of their marketing activity reveals careful and often extremely precise calculations of gaining optimal returns on resources controlled by the household economy. This same maximizing logic can be found in other areas of peasant behavior, revealing that priority is given to household interests over community goals.

An analysis of fertility in Yangbei before the recent strict birth control policy was implemented provides a clear example of household maximizing behavior that worked to undermine community goals. Yangbei peasants, like poor people elsewhere in the developing world, had large families out of choice.[16] Calculations of benefits and costs of having children were such that many peasants wanted to have more than two children, rather than less. In this respect, Yangbei peasants did not differ from other people in the decisiveness of economic factors in influencing fertility rates.[17]

To a large extent the problem was rooted in the system for allocation of grain to households. In the decades following collectivization, the amount of grain Yangbei peasants transferred to the state through taxation and compulsory sales left many households with less grain than they consumed each year. This compelled many households to buy grain on the free market at much higher prices than the state purchasing price for grain. For this reason, peasants preferred to receive payment for work in the collective sector in grain rather than cash, which purchased several times less grain than the cash equivalent in workpoints. However, the grain ration assigned to each household was based upon the number of people that belonged to the household. Thus with each additional child, a household gained a larger allotment of grain from the team. Moreover, a child's allotment, though smaller than that of an adult, was considered larger than what the child actually consumed. Thus households that had few children were at a disadvantage to those that had many in terms of basic grain allotment. For example, a one-child household might enjoy a high laborer/dependent ratio, but even if both parents worked hard for the collective and earned surplus workpoints above the cost of the grain allotted to them, the household often ran out of grain in the course of the year. It was thus forced to purchase expensive grain on the free market when surplus workpoints were converted to cash at the end of the year (*fenhong*) at prices several times higher than the state purchasing price. There was therefore strong incentive for the one-child family to continue to have children so that at least it could be paid in grain for work done in the collective sector. In addition, each child brought an additional private plot allotment and the possibility of having surplus grain to sell on the free market.

On the other side of the cost and benefit calculation, families with many children, even those with low laborer/dependent ratios, were not penalized for running deficit accounts with the team. In these households, the mother was likely to remain at home to rear small children. They therefore often had only one labor power, the husband, earning workpoints in the collective sector. As a result these households consumed more grain each year than they could pay for through accumulated workpoints, and thus they overdrew their acounts. But there was no penalty for overdrafting. Moreover, a household could overdraw for successive years on the understanding that at some future date it would repay the team for its deficit account. It received no cash income from the team during this period, since any surplus left in the account went to pay off the deficit accumulated over the years. In reality the household actually benefited from being able to overdraw. First, the real value of grain, reflected in the free-market price, was several times higher than the price set by the state, thus the household repaid the team at the state purchasing price for grain it could theoretically sell on the free market for much more. Moreover, the team charged no interest on the deficit account, nor did it set a schedule for repayment. Second, by overdrawing the household was, in a sense, subsidized by the team to have more children, further reducing the cost of having children. Some peasants calculated that when their children grew older and entered the work force, the children could repay the overdraft account in a matter of a few years. For this reason, they believed that it was better to have more children than to have to spend money buying free-market grain, since at least there was something to show for in the end, whereas grain was just eaten up.

Like fertility, the spacing of children was also influenced by economic considerations. While parents derived benefits from having children, they tried to reduce the costs as much as possible. One way to do this was to space the birth of children in such as way that the birth of the last child coincided with the maturation of the first child, who then could help in raising the younger siblings and reducing the amount the family must overdraw. A family with three children might decide to have a fourth, knowing that the first child was already old enough to babysit this younger sibling, allowing the mother to return to the fields to work. Soon the first child would also be old enough to earn workpoints to help pay for the cost of having additional children. If the family continued to have more children until they had five, by the time the third child had entered the work force the family would have cleared its debt to the team and begun to have a surplus account. Thus the birth and rearing of the last two children cost less to the household than the first three, and was in fact fully supported by the parents and three older children. The household was then in the enviable position of having a large grain allotment, a sizable private plot, and cash income from both the collective and private sectors.

Although the cost to a household of having more than two children was not high, and was even in some ways beneficial, the collective economy suffered if there were many households that overdrew their accounts. Such households were a burden to the collective economy in the sense that the team must use its reserves to subsidize them, leaving less funds for investment and saving. In poorer teams, the burden of supporting overdraft households had a demoralizing effect on the team as a whole. This was because poorer teams already had little surplus at the end of the year, and not only were they left with no funds for investment and savings, but they often did not have sufficient funds to pay households cash dividends. These households instead were given promissory notes that had no purchasing power.

Peasants wanted more than two children not only to maximize their household's share of grain and private plots; children also provided security in old age. Unlike urbanites, Chinese peasants do not have a social security program that guarantees income upon retirement. They therefore must depend upon adult children to support them in old age. Only the peasants unfortunate enough to be childless are compelled to rely on the team's welfare system. But because daughters move from their natal family to their husband's household, according to the custom of patrilocal marriage, peasants think of security in old age in terms of having sons. Thus parents with only daughters typically continue to have children with the hopes of giving birth to a son. Or they try to "adopt" a son-in-law into their household, but generally only the more prosperous households can hope to find a young peasant man willing to become an adopted son-in-law.

The incentives for individual households to have more than two children have led to a rapid increase in the village population. The population doubled by 1980 from its base in 1950, while the arable land actually declined due to road construction and land occupied by new housing. As a result, though production of grain increased over the years, per capita food consumption remained at the same level and possibly declined. Despite the realization that population growth contributed to stagnation in per capita food consumption and worked as a drag on further economic growth, it was not until the implementation of strict economic sanctions for additional children above the two-child norm that fertility began to decline rapidly, coming down to slightly over 1 percent, from over 2 percent prior to the implementation of a stricter government birth control policy. According to the new policy, households that did not comply with the birth control guidelines could not register additional births, and thus did not receive grain rations and private plots for children above the two-child limit. This effectively eliminated the economic incentives to have children under the collective farming system. However, it is likely that the return to household production will stimulate new pressures to have more than one or two children for peasant households that now feel the need for additional labor power. The optimal size of a peasant family for household production is above five, though over nine can result in inefficiencies.[18]

Another area where household maximizing behavior worked against collective interest involved the peasant men who ran deficit accounts in the team while they drew salaries from the state as cadres, workers, and government staff members. Because the households of state employees depended primarily on wives to earn workpoints to pay for household grain consumption, many customarily overdrew their team accounts. Rather than use a part of their salaries to pay off the overdraft accounts, some households of state employees preferred to defer payments until their children grew older and were able to pay back the team.

Lastly, a maximizing behavior characteristic of all households was to work at a slower pace on the collective fields. According to Yangbei peasants, the tendency for villagers to slow down in work for the collective sector was especially evident among women. There are a number of possible explanations for why women were apt to slow down while working for the collective economy. First, women married into the village, and, as strangers in the village, they were more likely to identify more narrowly with their husband's household rather than with the village as a whole. On the other hand, as a single-surname settlement, men, who unlike their wives were born and raised in the village, may have had a stronger identification with the team economy. Second, and more important, women in Yangbei had a longer work day than men. Not only did they work in the collective fields, but they also did most of the household chores. Caught in a "double bind," women's work in Yangbei was actually more strenuous, physically taxing, and required longer hours each day. It is not surprising that when she worked in the collective fields, a woman tried to conserve her energy for her household chores. When the team leader worked nearby she might work faster, but when he was out of sight she often slowed down and might even squat briefly when others were not looking. When women worked together as a group, an activist might scold those who worked more slowly, but there was an understanding among women that all benefited from working at a slower pace in order to have the energy to do the tasks they must accomplish for their households later in the day. What is pertinent to the argument is that women slowed down in the collective sector in order to be able to work more efficiently for their own households.

In conclusion, as the above examples of maximizing behavior of Yangbei peasants illustrate, the persistence of peasant household individualism resulted in household strategies that gave priority to household interests over collective interests. Even when peasants realized that all would gain through effective and productive cooperation, due to the "free rider" effect, peasants nonetheless gave priority to household private-sector work because such work accrued directly to the household. Likewise, peasants were surprisingly subtle in their calculation of household utility maximization, as evident in their marketing activity and their calculation of benefits and costs of having children.

## Collective Incentives and Group Pressure

It has been argued that collective incentives ought to provide adequate incentives to motivate peasants to work hard for their collective economy. According to this argument, peasants realize that their individual interests are bound to the welfare and productivity of the collective economy. The workpoint system provides differential rewards for individual effort. The collective, especially in a small production team or work group, moreover, exercises social control to pressure slower members to maintain production norms. Nonmaterial incentives, rooted in political study and consciousness raising, provide additional reinforcement to reward meritorious work.[19]

Though the workpoint system was supposed to reward differential output and quality of work performance, as in Unger's village, the actual range of workpoint distribution assigned to team members was relatively narrow. In most teams, workpoints were fixed only once a year. Some teams in Yangbei fixed workpoints as infrequently as once every other year. Thus workpoints tended to be relatively stable once assigned and did not measure seasonal or short-term changes in individual productivity. Moreover, in assigning workpoints, a team member's social status and seniority were taken into account in addition to the assessment of the quality of work input. Wives of brigade cadres, for example, were generally assigned higher workpoints, though they did not work harder than other women in the team. For these reasons the workpoint system at Yangbei tended not to be a very sensitive barometer of actual contribution on a day-to-day basis. However, the reforms introduced in 1979 and 1980 were supposed to tie collective incentives more closely to actual performance. But the rapid disintegration of collective farming shortly after the introduction of the reforms suggests that a more direct link between incentives and performance was not adequate to convince peasants of the viability of collective farming.

Nor was the problem rooted in an absence of effective social control over slackers or "free riders" in the team. As was evident in the group sanctions imposed on women who worked more slowly than others, the team did exert pressure on slackers to maintain production norms. Though occasionally a slacker might duck behind tall rice stems, or slow down when others were not looking, the greater part of a work day was spent working alongside other team members. Thus surveillance by team members of each other's actual work contribution was maintained. Moreover, a villager's social standing within the team, usually the size of a small hamlet or neighborhood in a hamlet, was to a large extent determined by his or her contribution to the team economy. In an informal reputation survey of team members, there was a close correspondence between a villager's social standing and the total workpoints earned in the course of a year. Hard work and skill were thus rewarded not simply through higher income from the collective sector, but also by higher social standing within the hamlet community. Similarly, those who contributed less, because they were seen as either lazy, slackers, or less competent, were typically those who were rated lowest among team members. These villagers were often the object of derisive teasing and contempt among fellow villagers.

American management studies have demonstrated that production norms can be and are maintained through group social pressure, especially in small work groups where face-to-face interaction is part of the work process.[20] But often there exists an informal understanding within the group to maintain production norms at a lower rate than what the work group could potentially sustain.[21] This was true of Yangbei women, as a means to resolve their "double bind" in favor of household work. Likewise, due to anxiety of "free riders" the stronger, more experienced, and more capable farmers, according to my informants, did not contribute their best effort. They complained that though they received the highest workpoint assignments in the team, this still did not fully compensate them for the value of their contribution. According to my educated youth informants, these experienced and capable farmers were among the most discontented members of the team, since they continued to believe that they could do better if they were to farm on their own. Thus social control appeared successful in maintaining a bottom-line production norm, but it apparently did not succeed in overcoming the effect of peasant household individualism and the "free rider" dilemma.

## History

During my field work in the village there was a tendency to idealize life there following land reform and prior to collectivization. My informants in Ithaca had told me that peasants described this period as one of prosperity and well-being. In life history interviews conducted in Yangbei, peasants recalled that per mu grain yields before collectivization were as high as the best years under collective farming, but more importantly per capita grain consumption was never higher. By contrast, peasants associated the years when the state pressed for rural radicalism, the Great Leap Forward and the period of brigade accounting in the early 1970s, with the leanest years, when subsistence margins were at their lowest for peasant households. At the time of my field work in 1980, food consumption had recovered from the slump caused by the latter part of the Cultural Revolution. Per capita grain consumption rose from 333 jin (unhusked rice) in 1977 to 600 jin in 1978, and it dropped again to 500 jin in 1979. This was during the transition to household production, when eleven teams were divided into twenty-one smaller teams. Yangbei experimented with various forms of small-group responsibility systems at this time. Yet, older peasants maintained that per capita food consumption was still lower than it had been before the Great Leap Forward. A number of peasants complained that household grain stores were still somewhat low. Despite the improvement in grain production, peasants continued to believe that the early 1950s were the best years, when household grain stores, meat consumption, and wine supply were plentiful even following the traditional celebration of the spring holiday, a week-long period of feasting and relaxation. Without documentation of actual per capita food consumption during the early 1950s, it was difficult to assess to what extent peasant assertions of better times were in fact accurate. But whether they were or not, Yangbei peasants believed that they were doing better when they farmed as individual households, and when cooperation was voluntary, as it was under the mutual aid teams from 1953 to 1955.

Collectivization swept through Yangbei in the winter of 1956. According to cadres who led the collectivization movement, Yangbei completed its transition to collective farming in a very rapid sequence of events. In a span of a few months, Yangbei moved from the large mutual aid teams directly to the higher level Agricultural Producer's Cooperative. Preparations were being made in the village in the spring of 1955 to establish lower level APCs, where land, draft animals, and farm implements were still legally owned by the household. But this first stage never got off the ground. Instead, during the winter slack season, cadres from the county government came to Yangbei, influenced by the national "socialist upsurge" mobilization, and pushed for setting up collectives at a faster pace than originally planned.

Though middle peasants, who had the most to lose in land, animals, and farm implements, were unhappy and reluctant to join, there was no active opposition to collectivization in Yangbei. According to life history interviews, peasants had enormous confidence in the leadership of the Chinese Communist Party. This grew out of Yangbei's long association with the Maoist revolution, as an early guerrilla base area in the Fujian-Jiangxi base area. Former guerrilla fighters and cadres in Yangbei formed a ready core of leadership for the new cooperative. The combined prestige of the village party members and the county cadres apparently was sufficient to mobilize enthusiasm for collective farming.

The harvest following collectivization continued the string of good harvests of the early 1950s. This suggested that the transition from the mutual aid teams to full-scale collectivization, though extremely rapid, was nonetheless quite smooth and did not disrupt agricultural production. This was all the more striking in light of the difficulties experienced in mobilizing enthusiasm and support for village-wide brigade accounting in the 1970s, as the higher level APC was a village-wide organization that employed a form of brigade accounting. Enthusiasm for collective farming probably peaked during the first year of the Great Leap Forward, as peasant belief in the credibility of the party leadership came under extreme strain. Wuping county was a center of the so-called communist wind in the Great Leap. In Yangbei, peasants were mobilized to build backyard iron furnaces, and they melted down cooking ware and pots, set up mass cooking halls, farmed in militarized work units, and were deployed by the newly established people's commune to build roadways far from the village. Although weather conditions were reported to be quite favorable, so many of the able-bodied peasants were drawn from agriculture to build roadways that not enough hands were left in the village during the critical planting and growing season.

The disruption caused to agricultural production by the Great Leap mobilization resulted in dramatically lower grain yields. Yet, the Xiangdong People's Commune reported unprecedented bumper harvests to the county government. It was on the basis of these exaggerated reports that the state quota for compulsory sales of grain was fixed. Before the mistake could be rectified, famine had already struck in Yangbei and Wuping county. During the difficult years after the failure of the Great Leap Forward, peasants received no support from the community or state to sustain them through the famine. The brigade had no savings or grain reserves. Peasants instead relied upon their own households to get them through these years. Reliance upon a household-based strategy for survival was reinforced by the state economic recovery policy, which permitted households to farm on their own under the household contract system and relaxed controls on the private sector and rural markets, the economic domain of the household economy after collectivization.

The failure of the Great Leap Forward was a shattering experience for Yangbei peasants, and it may have permanently colored their basic attitude toward radical forms of collective farming. It is impossible to know how long the enthusiasm—if it was as genuine as peasants recollected—that characterized the first years of collectivization could have been sustained had there not been such a massive failure. At any rate, the heroic period of collective farming in Yangbei, when peasants were willing to sacrifice household individualism for collectivist goals, proved short-lived.

## Economy of Scale—Household Versus Collective Production

Collectivization had two broad purposes: to bring about the rapid modernization of agriculture through insititutional development requiring a low rate of capital investment in agriculture, and to regularize the extraction of agricultural surpluses by the state to help finance socialist construction. It was predicated on the idea that the concentration of land and labor into larger units and the pooling of agricultural implements and draft animals would allow for fuller utilization of underutilized resources. It was thought that as a result, the productive forces suppressed under traditional arrangements could be fully unleashed, and economic growth accelerated. In reality, collectivization imposed large-scale organization on a small-scale economy. In Yangbei, there was a limit to efforts to concentrate the size of land plots due to topographical conditions of a mountainous area. Though less fragmented than before collectivization, the cultivable land was still divided into relatively small parcels, with larger plots in the alluvial valley, smaller plots on terraced land in the surrounding hills, and tiny plots scattered along mountainsides farther from the village settlement. Farmers continued to rely upon draft animals and labor-intensive means of cultivation. Even the widespread introduction of walking tractors in the mid-1970s did not change the labor-intensive cultivation that characterized rice agriculture in South China.

Whether economies of scale can be achieved by concentrating labor into larger production units in rice agriculture is still an open question. Concentration of labor may facilitate the mobilization of labor for infrastructure construction, such as roads, public buildings, and irrigation systems, but it is not clear that it actually leads to greater efficiency in wetland rice agriculture. As Fei Xiaotong wrote, "when work is mainly done by hands and feet, the advantage of division of work is reduced. Extensive organization in such enterprises gives no appreciable profit but rather complicates human relations."[22] In Yangbei, it was certainly evident that private plots were far more productive per area unit than were the collective fields. Economies of scale in agriculture are very limited, according to agricultural economist Alan Richards. "Even in heavily mechanized California, most economies of scale in irrigated field crops are achieved with farms of less than 200 acres. For relatively unmechanized peasant agriculture, there are few, if any, economies of scale."[23]

On the other hand, if peasant recollections of the larger household grain surplus of the early 1950s are accurate, collectivization probably resulted in a more effective system for extracting agricultural surpluses from the village through taxation and compulsory grain sales. Indeed, if there was one issue on which all Yangbei peasants could unite, it was in the various schemes to reduce the actual amount of grain the village must sell each year to the state. The quota for compulsory sales of grain restricted peasant grain consumption to levels closer to subsistence margins than peasants would have if they were not indirectly taxed through compulsory sales. Quotas also prevented peasants from setting aside land to grow cash crops that are more profitable than grain.

The reforms introduced in 1978 helped to establish more effective collective incentives based upon actual performance and reduced the size of production teams. These measures contributed to higher per mu grain yields and per capita food consumption.[24] The assumption guiding the 1978 reforms was most clearly articulated by Yangbei's party secretary. He remarked to me one day in explaining the reforms, ''because of the long tradition of individual household production, the closer we can approximate individual household production, while still maintaining the collective system of ownership, the higher the productivity we can expect to achieve.'' Given this assumption, I asked why Yangbei did not adopt the household contract system of the early 1960s, which allowed for household cultivation within the framework of collective ownership.

Team cadres explained to me that only the team could purchase chemical fertilizer, insecticides, and walking tractors, whereas individuals could not. Moreover, only the teams qualified for the low-interest credit extended by the state to purchase these modern inputs. They also pointed out the problem of using team-owned equipment and assets if individual household production were adopted. It would be a difficult and divisive process to decide who could, for example, be the first to use the team's walking tractor or oxen. Also, it was pointed out to me that households that were short on labor power and skilled farmers with sufficient breadth of experience to handle the entire cycle of agricultural production, such as female-headed households or households of cadres and workers who did off-farm work, would experience difficulty in adjusting to household production.

Still unsure of the permanency of the new pragmatic drift of state policy toward smaller units of production, more leeway for market forces, and more scope for material incentives, many peasants probably were too cautious to be assertively vocal about their preference for household production. In retrospect the party secretary's assessment of what it would take to unleash productivity reflected a keen appreciation for the strength of peasant preference for household production. Though I continued to ask questions to probe more into the reasons why some families wanted to return to household production, I never actually was able to interview peasants who freely discussed this conviction. But I could infer from statements made to me who some of these peasants might be—for example, the older peasants who emphasized the belief that things were much better in the early 1950s, or those who told me that despite recent improvements, life was still difficult, food stores low, and diet barely adequate. These peasants were telling me indirectly that they lacked confidence in the collective farming system and preferred the period of the early 1950s when farming was done according to individual households. Moreover, my informants in Ithaca had told me repeatedly that they were absolutely sure, on the basis of their knowledge of peasant attitudes, that Yangbei peasants wanted very much to return to household production. They claimed that few were willing to voice such views in the 1970s publicly, but in smaller circles of close friends, they talked more openly about their disillusionment with collective farming. My informants told me that all of the older peasants were still very clear about the boundaries of the land they farmed before collectivization, and they even enjoyed joking about this in public when peasants worked together in work groups.

In fact, my informants repeated claims about peasant preference for household farming were one of the primary reasons I felt it was necessary to go to Yangbei myself to conduct my own independent field work. I believed that my informants' statements of peasant preference for household farming reflected bias against collective farming. It is only with some humility that I confess now that my initial rejection of my informants' claims and evidence of household individualism reflected my own preconceptions. I had imagined early in my study that I would discover in a detailed ethnographic account the strength of communal bonds, first stemming from the natural bonds of a single-surname village and then being reinforced by two decades of collective farming. I anticipated finding a rich variety of informal institutions reflecting deeply ingrained cooperative behavior among peasants. In a sense, I very much wanted to find this, since my own preference was to discover that cooperative farming had deep roots in Chinese villages. Thus, even though my field work in Yangbei turned up sufficient evidence to confirm the claims of my informants in Ithaca, I still clung to my belief that somehow collectivized farming would persist through the reforms. A more perceptive view would have been to interpret the reforms, and the subtle statements from cadres and peasants pointing to the existence of preference for household farming, as a transitional stage in the dismantling of collective farming.

## Division of Labor

Despite changes in the technology of rice agriculture—the use of walking tractors, the improvement of seeds, and the utilization of chemical fertilizer and insecticides—the simple fact remained that the division of labor contained in a peasant household was still adequate for handling the entire cycle of agricultural production. This is especially the case in areas where the per capita size of land is small, resulting in an excess of labor power for agriculture. In Yangbei, most households contained an adequate division of labor to farm on their own. The median household size in the four production teams where I conducted household surveys was six members, with two labor power per household. Per capita arable land in Yangbei was about 1.4 mu, high for China, and much higher than along the coastal farming areas in Fujian. Yet, a 9-mu farm for a household of six to eight members was quite manageable within the context of the household division of labor. Female-headed households, households with low laborer/dependent ratios, and households where a husband lived away from home as state cadre, PLA soldier, or worker might be expected to have difficulties in farming alone. They probably would have continued to benefit from collective farming, especially since these households were more highly represented among the households with overdraft accounts. But these households constituted a minority of Yangbei households. I learned through interviews that the households that were discontented with collective farming tended to be those that had favorable laborer/dependent ratios and were headed by experienced and highly capable peasants. For these households, strong in labor power and led by skilled farmers, the belief that they could do better farming alone was quite strong, according to my informants. These households might be expected to take the lead when the opportunity arose to push strongly for individual household production, as they were among the most influential in their teams. But without knowing the precise events that led to the final break-up of collective farming in Yangbei in 1981, it is difficult to say what coalition of households pushed for a return to household farming, and which households opposed this, if there were any that did so actively. But recent field work conducted after the breakdown of collective farming, in villages that share similar conditions as Yangbei, emphasizes the view that there was a virtual upsurge from below in favor of individual household farming, once the state gave its approval.

If the analysis of the breakdown of collective farming in Yangbei is valid, in other areas in China where the peasant household is an adequate production unit it might be expected that peasants have already reverted to household farming of some form, or would prefer to do so. This would be especially true of areas where per capita arable land is small and there exists a surplus of agricultural labor power; in poorer, more backward areas, where agricultural mechanization has been slow to develop; areas with a tradition of peasant entrepreneurship and developed household sidelines; and also villages where collective farming never achieved levels of efficiency and productivity adequate to convince peasants that their basic interests were tied to the collective, more so than even their own households.

In the areas of China's countryside where the division of labor has developed to the point that households are no longer adequate as production units, however, peasants might be expected still to prefer collective farming over individual household production. The following areas are likely to have characteristics favorable to collective farming: those situated in rich agricultural areas close to urban centers where a part of the male labor force has been drawn away from agriculture by industry or construction projects; areas along major transportation systems that allow for easy access to urban markets, and where collective sideline industries are well developed and profitable; model and successful brigades that have a long history of success at collective farming, and which may receive subsidies and special inputs from the state based upon continued success in collective farming; or a combination of these circumstances which result in the development of the division of labor to the point that households can no longer function as adequate production units, or where peasants no longer depend upon their households for basic economic needs.

## Conclusion

This chapter has analyzed the sources of peasant preference for individual household production in Yangbei. As Theodore Schultz has written, "Much of our trouble in understanding agriculture in poor countries arises from misconceptions of the preferences of the people concerned and of the role of preferences in economic behavior."[25] In my analysis of the sources of peasant preferences, I have taken an "economic approach," which assumes that Chinese peasants respond to incentive structures as people do elsewhere in the world. Surprisingly, this point has often been overlooked in past analyses of the performance of Chinese agriculture and by the past policies of the Chinese government. Despite the growth of welfare services in collective agriculture, peasants still must rely upon their households for basic security and well-being. For this reason they are likely to identify their interests with those of their household, and subordinate individualistic behavior to allow the household to function efficiently as an economic unit. Within the collective economy, each household seeks to maximize the utility it derives from participating in it, as well as from the private sector that exists alongside the collective sector, and in competition with it. Hence, peasants are more likely to prefer household goals than individualistic and community goals (hypothesis 1). Peasant calculations of household utility maximization are quite subtle, often revealing a sophisticated investment logic, and are frequently at the cost of collective goals and interests. This becomes apparent when the household calculation of optimum gains conflicts with collective goals, such as in the case of birth control. In a sense each household is engaged in a zero-sum game in which the advantages it obtains for itself through household individualistic behavior are at the cost of other households in the same team. Yet, the fear of losing material utility to "free riders" tends to reinforce household individualism, lest the household lose out, as do those, for example, that fail to have children, and thereby risk not getting paid for surplus workpoints or having to purchase expensive grain on the free market.

The ability of the household to subordinate individualistic behavior to its function as an economic unit and the habit of maximizing household utility even for very marginal gains render the household a highly efficient production unit. As the party secretary observed, "the closer we can approximate individual household production while still maintaining the collective system of ownership, the higher the productivity we can expect to achieve." Because of the past mistakes and inefficiencies experienced in the collective sector, brigade and team farming failed to demonstrate to peasants that they were superior to what was achieved in individual household farming prior to collectivization. More importantly, as long as households continue to be adequate production units, capable of handling the full cycle of agricultural production, peasants are likely to prefer individual household production to collective forms of production (hypothesis 2). By farming alone, peasants gain the satisfaction that all of their work directly benefits their own household, whereas a persistent fear in collective farming was that by working harder than others they benefited their own household only indirectly, while supporting "free riders" who worked less hard or were less skilled as farmers.

An underlying implication of this argument is that Chinese peasants are willing to forsake the security provided by collective farming when given the opportunity to choose individual household production. However, in households that lack confidence in their capability as a sufficient production unit, and in richer, more developed localities where the division of labor has developed to the point that households cannot function as adequate production units, peasants are likely to prefer and sustain cooperative farming without the imposition of state power.[26]

# Notes

1. Benedict Stavis, *People's Communes and Rural Development in China* (Ithaca: Rural Development Committee, 1977); Kenneth Walker, *Planning in Chinese Agriculture*, (Chicago: Aldine, 1967); and Byung-joon Ahn, "The Political Economy of the People's Commune in China: Changes and Continuities," *Journal of Asian Studies* 34 (1975):631–58.

2. William L. Parish, "Socialism and the Chinese Peasant Family," *Journal of Asian Studies* 34 (1975):613–30.

3. See for example, William L. Parish and Martin King Whyte, *Village and Family in Contemporary China* (Chicago: University of Chicago Press, 1978).

4. James Scott, *The Moral Economy of the Peasant* (New Haven: Yale University Press, 1976). The view that peasants are risk averse is widespread in the literature.

5. Samuel L. Popkin, *The Rational Peasant* (Berkeley: University of California Press, 1979).

6. Theodore Schultz, *Transforming Traditional Agriculture* (New Haven: Yale University Press, 1964); see also Gary S. Becker, *The Economic Approach to Human Behavior* (Chicago: University of Chicago Press, 1976).

7. William W. Murdoch, *The Poverty of Nations: The Political Economy of Hunger and Population* (Baltimore: Johns Hopkins University Press, 1980) pp. 95–165.

8. Dennis L. Chinn, "Land Utilization and Productivity in Prewar Chinese Agriculture: Preconditions for Collectivization," *American Journal of Agricultural Economics* 59:559–64.

9. R. Albert Berry and William R. Cline, *Agrarian Structure and Productivity in Developing Countries* (Johns Hopkins University Press for the I.L.O., 1979), p. 194. I am grateful to Alan Richards, Department of Economics, University of California at Santa Cruz, for providing me with this reference.

10. George C. Homans, *The Nature of Social Sciences* (New York: Harcourt Brace & World, 1967), p. 22.

11. Since this chapter was written in July 1981, the household mode of production has become dominant in the Chinese countryside. Implementation of the household responsibility system grew into a nationwide campaign, and in some areas it was virtually imposed on peasants, often against the opposition of brigade cadres. We still do not have systematic data on why in some areas peasants welcomed the household responsibility system and in other areas there was opposition.

12. Parish, "Chinese Family Under Socialism."

13. Theda Skopol, *States and Social Revolution* (Cambridge: Cambridge University Press, 1979), pp.147–54; and Fei Hsiao-t'ung, "Peasantry and Gentry: An Interpretation of Chinese Social Structure and Its Changes," *American Journal of Sociology* 52, 1 (1946):1–17.

14. See Margery Wolf, *The House of Lim* (Englewood Cliffs: Prentice-Hall, 1968).

15. George Foster, "Peasant Society and the Image of Limited Good," *American Anthropologist* 67 (1965):193–315.

16. See George Casper Homans, *Social Behavior: Its Elementary Forms* (New York: Harcourt, Brace and World, 1961).

17. Gary S. Becker, *A Treatise on the Family* (Cambridge: Harvard University Press, 1981); Theodore Schultz, *Economics of the Family: Marriage, Children, and Human Capital* (Chicago: University of Chicago Press, 1975); and John Caldwell, "The Economic Rationality of High Fertility: An Investigation Illustrated with Nigerian Survey Data," *Population Studies* 31 (1977):9.

18. William Murdoch, *The Poverty of Nations*, pp. 140–51.

19. See Carl Riskin, "Maoism and Motivation: Work Incentives in China," in

*China's Uninterrupted Revolution*, ed. Victor Nee and James Peck (New York: Pantheon Books, 1975), pp.415–62; Dennis L. Chinn, "Diligence and Laxness in Chinese Agricultural Production Teams," *Journal of Development Economics* 7 (1980):331–44; and Dennis L. Chinn, "Team Cohesion and Collective Labor Supply in Chinese Agriculture," *Journal of Comparative Economics* 3 (1979):375–94.

20. Fritz J. Roethlisberger and William J. Dickson, *Management and the Worker*, (Cambridge: Harvard University Press, 1939); George Casper Homans, *The Human Group* (New York: Harcourt, Brace and World, 1950).

21. Homans, *Social Behavior*, pp. 48–154.

22. Fei Hsiao-t'ung, "Peasantry and Gentry," p. 14.

23. Personal communication from Alan Richards, February 23, 1982.

24. See Victor Nee, "Post-Mao Changes in a South China Production Brigade," *Bulletin of Concerned Asian Scholars* 13 (1981):32–40.

25. Theodore Schultz, *Transforming Traditional Agriculture*, p. 28

26. In many areas cooperatives have been formed, owned, and run by small groups of three to five households. These new cooperatives may become a popular form of joint-investment vehicles for peasants to go beyond the limitations of household production. They appear to be small business enterprises in which capital and labor is pooled for greater productivity and profits. Whether such corporate economic units will be allowed to develop in scale and importance is still uncertain. See *Renmin ribao*, January 20, 1984, p.2.

# PART THREE

## New Patterns of Equality and Inequality

# 8

# Income Inequality and the State
# Mark Selden

A substantial literature credits the People's Republic of China with impressive gains in the reduction of inequality. Dwight Perkins, for example, observed that China has "clearly reduced intra-village income differentials in a major way; per capita differences of 2:1 from the richest to the poorest family are probably rare. . . ."[1] Similarly, Alexander Eckstein noted evidence both of "a compression of average urban-rural income differentials" and "narrowing of income differentials . . . in the inter-regional distribution of income."[2] Yet, at least until the early eighties, quantitative information about changing local, regional, and national patterns of income and inequality was at best scarce and highly selective. In light of the extensive development literature elucidating the tendency for both interhousehold and spatial inequalities to widen during early phases of development, and the extensive praise for Chinese reduction of inequality, with new data we can now more precisely gauge that performance.[3]

This chapter introduces micro-level data, reexamines interpretive frameworks, and proposes conceptual approaches toward the analysis of Chinese rural income and inequality in the years 1949–1979. Discussion centers on two important components of rural income inequality and the impact of state policies and institutional change for redressing—or aggravating—them. These are local inequality, or the differentials among households in the same community or income pooling unit; and spatial or geographic inequality among units, localities, and regions.

Marc Blecher, Chiu Cheng-chang, Christopher Peck, Peter Nolan, Thomas Rawski, Benedict Stavis, and particularly William Parish generously provided critical suggestions, data, references, and technical assistance. Their contributions are gratefully acknowledged.

Income is an imperfect measure of inequality, and a full treatment of inequality issues would require analysis of many other factors, including differential access to power, nonmonetary sources of prestige and consumption, and the redistributive impact of state and community welfare systems. Nevertheless, even in a society such as China's, in which commodity production remains at a low level, substantial interchange takes place outside the market, and the roles of state and collective are powerful, income distribution provides one vital and measurable gauge of inequality and of the overall performance of the system.

## Local Income Inequality

Wugong is a prosperous model village that has had considerable success in collective agriculture. Its per capita income places it among the richest villages in China. This is a position attained in part as a result of its early cooperative achievements, in part due to the fact that it was selected as a model village in the 1950s and has since made judicious use of state subsidies and a number of other advantages. It is located in southeastern Hebei province in the poor Heilonggang region on the North China plain. In 1977, it had a population of 2,536, farming 243 hectares of land that were mostly in wheat. For administrative purposes, the total village constituted a single production brigade, and this brigade was then divided into three production teams, which were the effective units for collective farming and income sharing.

We have household survey data on income in two of these subunits. The most prosperous team in the village was team three, consisting in 1977 of 998 people in 232 households. In 1977 team three's average per capita income from all sources was 245 yuan and from collective sources 214 yuan. Team two was the poorest. Its 1977 per capita collective income was 137 yuan. Both did well by national standards, ranking comfortably within the top 10 percent of all teams, yet distinct differences in economic performance and income differentiated the two teams.[4] The per capita collective distributed income of team three exceeded that of its neighbors in team two by 56 percent despite comparable land endowments and the equalizing effects of income generated by a substantial brigade-level economy.

### *Sources of Local Inequality*

With land reform in 1947 in this area, and then with the collectivization of agriculture in the mid-1950s, major sources of income inequality in the village were removed. Nevertheless, even with the elimination of all significant property-based inequalities, as well as the loss of diverse off-farm sideline and commercial activities, some bases of income differentiation remain within the collective sector. Close analysis of the data reveals that the most significant of these was simply the number of able-bodied laborers one had in one's household. In Wugong as elsewhere in China after 1956, the collective farm (usually the team) continued to pay people in cash and in kind on the basis of workpoints according to how much each family member worked. In contrast to many other places, however, in Wugong these workpoints were only minimally differentiated by age, sex, or skill. While most men earned ten workpoints a day, in the late seventies women often earned nearly as much, getting nine to ten workpoints a day. Women earned fewer total workpoints each year primarily because they shouldered the main responsibility for family chores and could not work in the collective fields as much as men. Older men also did well in comparison to younger men, continuing to earn almost as much as younger men long after their strength had declined, so long as they continued to report for work in collective endeavors. All of these features contributed to narrowing income differentials among participants in the collective economy of the village.

Within each team one can nevertheless discern significant inter-household income differentials. The principal variable is the ratio of income earners to total household population. Households with more children and retired parents to support consistently ranked among the lowest in per capita incomes. The most prosperous households were those with children in the 15–28 age bracket. Typically, these young people had finished school and joined the work force but continued to live at home and contribute their income to the household before becoming married. Another group that did surprisingly well was households comprising old people with no dependents left to support. There were twenty-one households in team three with two people over age sixty and no one else to support. These twenty-one households received an average per capita income of 289 yuan per year, well above the team average of only 245 yuan.

Conversely, those village households that fared most poorly had many dependents and few able-bodied laborers. Households with many young, pre-school-age children are doubly hurt when the mother must stay home to tend them, thereby forfeiting her workpoints from the collective farm. (This village had no nursery in 1977.) Old people who cannot work but live alone also obviously have low incomes. The poorest person in team three was a woman in her seventies who lived alone and received a total annual income of only 37 yuan.[5] A primary source of income differentials in the village then is simply one's place in the family life cycle—what the Russian agronomist Chayanov referred to as demographic differentiation. Those at the beginning and end of this life cycle suffer the most—those near the beginning have too many young children and sometimes dependent parents or grandparents as well; those near the end may be able to earn no income of their own and must fall back on the support of their children and, in the last resort, on collective welfare payments. Those in the early-middle years of the life cycle with many able-bodied laborers still at home do the best. The cyclical character of this income pattern distinguishes contemporary rural China from many other developing nations in which major intravillage income differentials are the product of highly unequal access to land and other means of production.

The search for sources of income inequality in Wugong produces surprisingly few other important differentiating factors. Service in local village leadership and administrative roles produces no apparent income advantage. In team three in 1977 nineteen households had one or more members active as team leader, brigade secretary, commune leader, or external army officer. Among these nineteen households, per capita income averaged only 227 yuan, well below the team average of 245 yuan.

Another group that fared more poorly than expected was households with members working outside the village in state-sector jobs. In team three, twenty-five households reported remittances of state salaries from parents or children working outside the village. They worked as military and state officials as well as technical and administrative specialists, teachers, and factory workers. These posts are coveted by village youth, and villagers with more education and political connections have a better chance of getting them. State-sector jobs bring secure monthly salaries with important fringe benefits, such as retirement pensions and health insurance, attached. Nevertheless, assuming that the villagers were forthright in reporting remittances from these kin when we conducted the household survey, in most cases the household back in Wugong village did not reap great financial rewards from having jobholders in the state sector. The twenty-five team three households with state jobholders had average per capita incomes of only 238 yuan, slightly below the village average of 245 yuan. And unlike some other villages in which substantial income came from contract and temporary industrial labor, or from private commercial activities, such income sources were absent in our data.

An example of one person who has been a leader and who has kin in state jobs illustrates the dominance of household life cycle position in incomes. Zhang Duan was the village party secretary for fifteen years before retiring in 1974 as a result of crippling illness. He heads a nine-member household that includes three party members and a PLA officer stationed in Beijing. A daughter-in-law holds the position of commune vice secretary and a son works in a county textile factory. This household well illustrates how family connections can lead to better jobs. But at the time of our survey the family back in the vilage reaped few direct benefits in the form of income. With three young dependent children and neither Zhang Duan nor his wife earning income, the reported per capita income was only 82 yuan—one of the lowest in the team.

What do these income differences signify for prosperous and poorer households in a model village? The household survey discloses that as of 1978 the largest differences in consumption patterns centered on the size and quality of housing and possession of five consumer durables: bicycles, sewing machines, watches, clocks, and radios. Comparing the top and bottom deciles of per capita incomes, we find that by these measures households in the top decile were twice as well off as those in the bottom decile. The richest decile had 1.5 rooms and 1.2 items of the five consumer durables per person, whereas households in the poorest decile averaged just .77 rooms and .61 consumer durables per person. Other important differences in consumption patterns centered on the scale and style of weddings, funerals, and other important life cycle moments, as well as a more varied diet and superior clothing. Income differences in Wugong do not significantly affect issues of subsistence (due to effective collective welfare policies), nor do they substantially affect the life opportunities for the next generation by assuring superior access to education, preferred jobs, or the ability to migrate to the cities.

Two other points should be noted. In poor villages, remittances from kin working outside the village may make a more significant difference in household incomes. But in this prosperous village, remittances from outside workers, including those on state payrolls, have not changed local income distribution that much. From the perspective of the great majority of villages with little land and abundant labor power, however, it is essential to place as many individuals as possible in jobs that produce incomes from the state and from other external sources.

The same point can be made about work in the private sector, including work on private plots and the selling of household handicrafts in peasant free markets. Both private income and salary remittances constituted only a small part of the income of team three in the late 1970s—13 percent, divided about equally between private sector activities and remittances. This was a low figure compared to national figures at that time, and far below national figures in the eighties—in 1981 private-sector earnings constituted 38 percent and remittances and other miscellaneous earnings 10 percent of peasant incomes throughout China.[6] Nevertheless, the same principles probably hold. Households that send their members to work in the private sector or in state-sector jobs outside the village will have fewer members left to work on the collective farm. Thus the increase in alternate earnings will in part be offset by reduced collective farm earnings, notably in units with substantial collective earnings.

Table 8.1 records the distribution of income in team three according to a number of measures, including the proportion of village income earned by various income groups, the income ratio of the richest households compared to the poorest, and an overall measure of income inequality called a gini coefficient—a larger gini coefficient implies more inequality. This table shows, not surprisingly, that the distribution of total income per household is less equal than the distribution of per capita incomes. Some large households appear to be rich because they have more members at work. But because they must share this income among more dependents, many such households have relatively low per capita incomes.

Comparing the richest and poorest 10 percent of households shows the importance of the demographic factor in the pattern of per capita incomes in Wugong's team three. The top 10 percent of households with mean per capita incomes of 427 yuan averaged just 2.8 household members. The bottom decile, which averaged 107 yuan per capita income, had an average of 4.5 members per household. To reiterate, the data strongly underline the cyclical pattern of income differentiation within the village in which neither sector (state or collective) nor occupation (agriculture, sideline industry, factory, cadre, or administrator) substantially affects household and per capita household income.

Table 8.1 shows that in this affluent team, extracollective sources of income have tended to moderate rather than exacerbate income inequality. It appears, then, that in Wugong households do trade collective for off-farm and private income and that the overall effect of these alternate sources of income is either neutral or slightly equalizing in its consequences.

Table 8.1

**Per Capita and Household Income Distribution in Team Three, Wugong Village, 1977 (percentages)**

| Indicator | Household collective income | Household total income | Per capita collective income | Per capita total income |
|---|---|---|---|---|
| Poorest 10% | 2.6 | 3.0 | 3.4 | 4.3 |
| Richest 10% | 19.0 | 18.7 | 17.8 | 17.2 |
| Poorest 20% | 7.3 | 8.1 | 9.3 | 10.9 |
| Richest 20% | 34.5 | 33.5 | 31.8 | 30.7 |
| Middle 60% | 58.2 | 58.4 | 58.9 | 58.4 |
| Poorest 40% | 21.1 | 22.7 | 25.3 | 27.3 |
| Decile ratio (richest: poorest 10%) | 7.3:1 | 6.2:1 | 5.2:1 | 4.0:1 |
| Gini coefficient | 0.27 | 0.25 | 0.22 | 0.19 |

*Source:* 1978 household survey of Team Three, Wugong Brigade.

Many of these patterns, with respect to the demographic cycle and the equalizing effects of extracollective income, are not restricted to this one rich team in a model village. Data on an average-income team in Guangdong province show a comparable distribution. With an average per capita collective income of 71.6 yuan in 1978, Team 12, Xintang Brigade, Tangtang Commune was just below the national average of 74 yuan in that year.[7] Alternate private and off-farm sources of income constituted only 12 percent of household income, much as in Wugong. Table 8.2 displays the types of incomes just as in the previous table, only this time the overall measure of inequality is not the gini coefficient but the coefficient of variation, which behaves much the same way. By this measure, income inequality again declines steadily as one moves from left to right across the table, from household collective income to per capita total income. Or, in other words, in both Wugong and this Guangdong team, alternate sources of income outside the collective sector moderate rather than exacerbate intrateam income inequality.[8] Indeed, the lion's share of household collective income inequality is the product of the demographic cycle. In this Guangdong village the top decile of households with mean collective incomes of approximately 130 yuan averaged 1.7 persons per household compared with the bottom decile with mean collective incomes of 45 yuan and 5.3 persons per household. More significant still, where the ratio of labor power to household size for the top decile was 1:1.0, for the bottom

Table 8.2

## Income Distributions in a Rich and Average Team
## (coefficients of variation)

| Indicator | Household collective income | Household total income | Per capita collective income | Per capita total income |
|---|---|---|---|---|
| Rich team (Wugong, 1977) | .47 | .43 | .37 | .35 |
| Average team (Guangdong, 1978) | .50 | .44 | .43 | .33 |

*Sources*: "Rich,"; Wugong data for 1977 from our survey. "Average team" data from 1978 report on 27 out of the 28 households in team 12, Xintang Brigade, Tangtang Commune, Guangdong Province, as reported in Keith Griffin and Ashwani Saith, "The Pattern of Income Equality in Rural China," *Oxford Economic Papers* 34 (1982), p. 203.

*Note*: The coefficient of variation is the standard deviation divided by the mean income for each category of income distribution. It provides a measure of inequality analogous to the gini coefficient. Like the gini coefficient, a larger coefficient indicates greater inequality.

decile it was 1:4.0. In other words, the households with the lowest collective incomes had the largest number of mouths to feed per labor power.[9]

The Wugong and Xintang cases illustrate the proposition that the primary determinant of one's income relative to that of neighbors within the team is the stage in the household life cycle, with the ratio between labor power and dependents providing the crucial indicator of per capita household incomes. Stated differently, a substantial portion of the inequality witnessed at any one point in time is a transient inequality that will even out in the course of the life cycle as children become mature laborers earning good incomes and later marry and form their own households. When people become too old to support themselves, they must rely on their sons for support, and in the last resort on village welfare programs. In poorer villages the downswings in a household life cycle have been further cushioned by the collective (team), which would allow households to draw grain for subsistence even when they could not pay for that grain out of current income. The household thus incurred a debt to the team to support current consumption, a debt which in some cases might never be repaid. This leveling out of the effects of the household life cycle suggests that we should consider lifetime inequality rather than income inequality at a single point in time. And by such a standard of intra-village and particularly intra-team lifetime equality rural China appears to be very equal.

## Comparisons of Local Inequality

Comparative data on villages in China and elsewhere helps place income inequality in Wugong in context. For China itself, data on local inequality within a single collective unit exist not only for the two teams in Wugong and for Xintang but also for six other collective units at different times and in different places. This sample includes prosperous and middle-income communities located both in wealthy and poor, mountain, plain, and suburban regions in six provinces in North, Northwest, Southeast, and Southwest China. Like most of the available data on rural China, the sample is nevertheless skewed toward well-to-do communities. In the total of nine units, only Lujiafu in Shandong, Fenghuan in Sichuan, and Xintang in Guangdong (discussed above) were close to the national average for rural incomes. All other units ranked in the top 20 percent of rural household incomes at the time they were surveyed. Several, including Wugong and Liulin, were exceptionally prosperous units in poor regions. In particular, this sample lacks data from hard-core poverty villages and from villages in frontier and minority areas.

Data are most consistently available on collectively distributed income—that is, income paid to each household by the collective (team) and excluding additional income from private endeavors or from work off the farm. The comparative data presented for eight units are not for per capita total earnings but for household earnings distributed by the collective (table 8.3). By our measures of inequality, the two teams in the village of Wugong are quite similar in most respects (columns 7, 8). Although the income ratio between the richest and poorest group differs significantly in teams two and three, the overall degree of inequality captured in the gini coefficient is about the same in both units. The same is also true of Liulin and Xishan in earlier years. Though differing in some specifics, all four units have degrees of local inequality that are in a similar range with gini coefficients ranging from 0.26 to 0.29. A group of villages in our sample exhibits substantially more equal income distributions than those discussed above. They include the middle-income Shandong village, Lujiafu, and the two relatively prosperous Guangdong villages of Dongguan and Dianbai with gini coefficients ranging between 0.16 and 0.17. The data now available cannot sustain the hypothesis that higher incomes correlate with greater intracollective income inequality (or the reverse). The sample suggests a plausible range within which estimates for income inequality in most Chinese communities may be expected to fall. Despite the small size of

Table 8.3

## Collective Household Income Distribution in Selected Chinese Villages, 1955 to 1979 (percent)

| Indicator | (1) Xishan Beijing, 1955 | (2) Lujiafu Shandong, 1958 | (3) Liulin Shaanxi, 1963 | (4) Dongguan Guangdong, 1974 | (5) Dianbai Guangdong, 1974 | (6) Fenghuan Team Two Sichuan, 1977 | (7) Wugong Team Three Hebei, 1977 | (8) Wugong Team Two 1978 |
|---|---|---|---|---|---|---|---|---|
| Poorest 10% | 3 | 6 | 3 | 7 | 6 | 2 | 3 | 2 |
| Richest 10% | 19 | 16 | 24 | 16 | 16 | 21 | 19 | 21 |
| Poorest 10% | 10 | 13 | 7 | 14 | 14 | 7 | 7 | 7 |
| Richest 20% | 35 | 30 | 37 | 29 | 29 | 37 | 35 | 35 |
| Middle 60% | 55 | 57 | 56 | 57 | 57 | 56 | 58 | 58 |
| Poorest 40% | 24 | 29 | 22 | 20 | 31 | 18 | 21 | 22 |
| Decile ratio | 6.3:1 | 2.7:1 | 8.0:1 | 2.3:1 | 2.7:1 | 10.5:1 | 6.3:1 | 10:1 |
| Gini coefficient | 0.26 | 0.17 | 0.29 | 0.17 | 0.16 | 0.31 | 0.27 | 0.27 |
| Per capita collective income | 138 | 62 | 126 | 157 | c.150 | 72 | 216 | 130 |

*Sources*: Columns 1–5, Marc Blecher, "Income Distribution in Small Chinese Rural Communities," *China Quarterly* 68 (1976):797–816. Column 6, Keith Griffin and Kimberley Griffin, "Institutional Change and Income Distribution in the Chinese Countryside," presented at the Conference on Development and Distribution in China, Hong Kong, March 14–17, 1983. Column 7, author's field research. Column 8, provided by Peter Nolan.

*Notes*: Xishan—a sample of 26 households in an advanced cooperative. Lujiafu—all 243 households with a per capita collective income of 62 yuan. Liulin—46 out of 50 households in two teams. Dongguan—a sample of 22 out of 98 households in a team. Dianbai—all households in a team. Fenghuan—31 out of 32 households. Wugong—all 232 households in team three and all 216 households in team two.

the sample, the internal consistency of the data, and its mesh with what we know about rural income structures, suggests that it provides a working guide to the range for income distribution in many Chinese rural communities in the era of collectivization from 1955 through the 1970s.

That even the most unequal villages in the sample are quite equal is indicated by comparison with income distribution data on Indian villages. Most available data from other countries aggregate local with regional inequalities into an overall measure of rural inequality. Such aggregate data preclude comparisons with the kinds of intravillage inequality data presented on China to this point. However, one study of eighty-four Indian villages does provide data on inequality structures within villages. In that study, the average gini coefficient per village was 0.46, which is well above the 0.16–0.27 range observed in the China sample, indicative of much higher intravillage inequality in the Indian case.[10]

There is reason to believe that most of the Indian differentials, and those in most other developing Third World capitalist nations, far from being overcome in the course of the life cycle, exhibit tendencies to remain or even to become exacerbated. These income inequalities are rooted above all in sharply differentiated access to ownership and control of land, which is frequently the primary source of employment, subsistence, and income.[11] By contrast, China's land reform, followed by the collectivization of land, large animals, and equipment, substantially reduced intravillage income inequality, and that pattern has persisted at least into the early eighties.

## Spatial Inequality

To this point we have only examined income inequality within single villages or collective units. But in addition to this type of income inequality, major spatial inequalities exist among Chinese villages and regions. These have persisted and in some cases increased. The major efforts at rural transformation in the 1940s and the 1950s—land reform and collectivization—left essentially untouched large spatial inequalities among villages and regions that were rooted in ecological, demographic, and socioeconomic differences. In spite of this limitation, the changes associated with land reform did have a significant impact on incomes throughout the countryside. In a study comparing household income in the early 1930s and 1952, Charles Roll found that the poorest peasants, the bottom 20 percent of households, almost doubled their income share from 6.0 to 11.3 percent between the 1930s and early 1950s (table 8.4). The gini coefficient measuring overall inequality decreased significantly. Equalizing resources within each village, then, had a significant impact on overall rural inequality. Roll concluded that 75 percent of the differences in per capita crop income were the product of interregional differences.[12]

---

Table 8.4

**Household Per Capita Rural Income Distribution in China, 1934 and 1952**

|  | Percentage share 1930s | Percentage share 1952 | Change in share |
|---|---|---|---|
| Richest 10% | 24.4 | 21.6 | −2.8 |
| Richest 20% | 42.0 | 35.0 | −7.0 |
| Upper-middle 20% | 23.9 | 21.3 | −2.6 |
| Middle 20% | 14.9 | 17.4 | +2.5 |
| Lower-middle 20% | 13.2 | 15.0 | +1.8 |
| Poorest 20% | 6.0 | 11.3 | +5.3 |
| Poorest 10% | 2.5 | 5.1 | +2.0 |
| Decile ratio | 9.8:1 | 4.2:1 | |
| Gini coefficient | 0.33 | 0.22 | |

*Source*: Charles Roll, *The Distribution of Rural Incomes in China: A Comparison of the 1930s and 1950s* (New York: Garland, 1980), p. 76; derived from Li Chengrui, *Zhonghua renmin gongheguo nongyeshui shigao* (Draft History of the Agricultural Tax in the People's Republic of China) (Beijing: Finance Publishing House, 1959), pp. 60-63.

---

The reductions in income inequality were largely restricted to changes within single villages. Spatial inequalities remained. Following Li Chengrui's classification of China into six regions by county per-capita agricultural incomes, it is evident that in 1952 following land reform, former rich peasants in the poorest counties still earned less than poor peasants in middle-income counties, despite the fact that in the poor regions per capita rich-peasant income was still twice that of the poor peasants.[13] (The class categories are based on land ownership and income *prior* to land reform.) Conversely, poor-peasant per capita incomes in the richest counties (group one) were substantially higher than rich-peasant incomes in the poorest counties (groups four, five, and six). Average per capita incomes in the most prosperous counties were five times higher than those in the poorest. On the basis of Li's data, we can estimate 1952 per capita income of the top 10 percent of counties at 700–750 kilograms of unprocessed grain and the bottom 10 percent at 150–160 kilograms, giving a decile ratio in the range of 4.4–5.0:1. In short, some of the largest rural income inequalities remaining after land reform can be accounted for in spatial or geographical terms, in this case using the county as a unit.

These sorts of spatial differentials are shaped in large measure by ecological factors, including differential terrain, climate, soil quality, precipitation, and access to water, transportation, and urban markets. They are also conditioned by human and social factors shaping local and regional developmental processes. Spatial differentiation in contemporary China takes the form of substantial differences in per capita income and opportunity among teams, brigades, communes, counties, and provinces; among mountain, plains, and suburban areas; between coastal and inland, industrialized and nonindustrial, Han and minority regions; and between city and countryside as well as between workers and peasants.[14] How have state policies impinged on these spatial factors?

Nicholas Lardy, who has conducted the most exhaustive analysis of China's regional inequality and redistributive policies, concluded that "the initiation of sustained growth of per capita GNP was accompanied by a simultaneous reduction of regional inequality."[15] Lardy documents substantial reductions of interprovincial inequality in domestic product and levels of industrialization. This result, in impressive contrast with the performance of most other nations in early stages of development, he tells us, was achieved by means of a leadership choice "to sacrifice some economic growth in return for achieving improved regional economic balance."[16] But whatever the validity of these conclusions, they rest on two kinds of data: provincial industrial output and interprovincial financial transfers. Lardy presents no significant evidence of reduced urban-rural or intrarural inequality.

Following Mao's 1956 discussion of "The Ten Major Relationships" and his calls during the Great Leap Forward and Cultural Revolution to eliminate "the three great differences," the Chinese leadership has repeatedly called for narrowing the urban-rural gap on behalf of the peasantry. Moreover, in maintaining low industrial wages, in shifting the terms of trade to the advantage of the countryside by raising state purchasing prices of agricultural commodities, and in diverse other ways, the Chinese state has periodically sought to narrow the gap.

Yet our attention is riveted on powerful countertendencies which have preserved or even exacerbated urban-rural and intrarural differentials at the level of both state policies and environmental constraints. Among the most striking features of Chinese rural development has been the official stress on self-reliance, with its favorable implications for units and regions that enjoy natural advantages such as terrain or access to water, sociopolitical advantages such as location in early liberated areas, technological-economic advantages such as a legacy of highly skilled agricultural and handicraft producers, and locational advantages such as proximity to cities or railroads.

A related approach, the emphasis on successful models, has also tended both to channel disproportionate state resources to a very small number of high-flying units and to serve as justification for not providing extensive subsidies to lagging units and poorer areas. Similarly, the transfer of 16 million educated youth to rural areas from the mid-sixties on balance increased the burden for the already overpopulated countryside. Stated differently, the Chinese state, which has taken seriously its obligations to prevent starvation in very poor or disaster areas, has rejected certain obvious redistributive methods, including progressive taxation and direct subsidies to accelerate development in poor regions. Quite the contrary. China's tax policies have been consistently regressive, stimulating rapidly developing units while acting as a brake on the economic performance of those who lag behind. China has not developed social security or national health systems that might work to the advantage of poorer areas, nor for the most part has it earmarked special development funds for chronic poverty regions. And in imposing high rates of accumulation on the countryside, in siphoning off much of that accumulation to accelerate industrial growth, and in curbing such important sources of rural income as those associated with traditional sidelines, handicrafts, and marketing, the state has maintained a series of policies that have favored the cities to the detriment of the countryside. Finally, when strategies emphasizing self-reliance and successful models are coupled with the formidable array of population controls which bar migration from poorer to more prosperous regions and from countryside to city, the result is not only to perpetuate but to reinforce spatial inequalities.[17]

Suzanne Paine has observed that population controls can cut both ways: While the prevention of large-scale migration out of depressed areas tends to perpetuate spatial inequalities, the same policies prevent the drain of skilled labor from poorer regions, to which we may add the energies of youth and those with entrepreneurial and other creative impulses.[18] She concludes that on balance the advantages in retaining skilled labor are decisive, and that compared with economies that permit free population movement, spatial inequalities will be relatively small. Population controls undoubtedly help to prevent the formation of the urban slums with their monumental problems of unemployment and marginalization common to most Third World nations. But under Chinese conditions of substantial and, at least until the late seventies, most rapidly growing rural labor surplus in poorer rural regions, the principal effect of population control policies on rural income has been to reinforce intrarural and urban-rural spatial inequalities. Whereas more prosperous regions are in a favorable position to accumulate and absorb surplus labor in a wide range of sideline and industrial enterprises, many poorer regions lag behind and find it increasingly difficult to employ the growing numbers in the labor force. This pattern has been reinforced by such state policies as ''taking grain as the key link'' (to the detriment of animal husbandry, forestry, and commerce), which hit hill and mountain regions particularly hard.

The most important and least studied structural limits on spatial mobility since the early fifties have been the restriction of peasants to jobs and residence within their teams through a tightly controlled system of work and travel passes and rationing by *hukou* (official residence) of food, cloth, and other necessities. To state the matter starkly, the majority of Chinese people since the mid-fifties have been legally bound by the state for life to residence and collective labor within tiny production units, teams, typically comprising thirty to forty households.[19] The weight of this fact, properly understood, shapes much that is distinctive of Chinese policy and practice, including issues of spatial inequality. Rural population control reinforces the fact that each cell, each team, brigade, and commune which comprise the rural milieu, is a community of destiny that structures the life opportunities and incomes of member households and individuals and, if this argument is correct, works to the disadvantage of poorer and more remote communities.[20]

Chinese national survey data released in 1981 present a clear view of distribution of rural poverty and spatial inequality by county in the late seventies. In 1977, 515 counties had per capita collective distributed incomes below 50 yuan, the officially designated poverty line. And even with the 1979 bumper harvest and increased state purchasing prices, 221 of China's 2,300 counties, altogether 88 million people, had average per capita collective distributed incomes below 50 yuan during all three of the years 1977–79. These poverty-stricken counties, scattered throughout China, cluster heavily in hard-core poverty zones, including 71 in the low-lying saline-alkaline areas of the North China plain, 70 in the Yunnan-Guizhou plateau and mountain regions, and 48 in the dry Northwest loess plateau. Sichuan, which pioneered in expanding the household sector, reportedly led the way in the reduction of poverty counties with per capita collective incomes below 50 yuan from 39 in 1977 to just 3 in 1979. Large gains were also reported by a number of North-China-plain provinces: Shandong reduced poverty counties from 63 to 26; Hebei from 51 to 13. Others, however, registered little change. In Gansu the number of poverty counties barely declined from 35 to 32 and in Guizhou it actually increased from 52 to 53.[21] The evidence underlines the fact that the national development strategy, which combined large-scale collectivized agriculture with heavy accumulation directed to heavy industry, bypassed large areas of the countryside in terms of the capacity to raise incomes.

## Combined Rural Inequality

We need not be content with just speculating about the consequences of persistent poverty in some areas combined with growing prosperity in others such as in rich delta plains and on the outskirts of major cities. In 1979 China's Ministry of Agriculture surveyed over five million rural collective units, or almost all of China's rural production teams. The published survey shows the distribution of average per capita collective incomes in these units. These data on spatial inequality can be combined with our data on local inequality to provide estimates of combined rural inequality throughout the nation. The methodological note at the end of this chapter details how these estimates were made.

For the purpose of these estimates, we assume that the distribution of private income, including remittances from urban kinsmen, is a more or less constant percentage of collective income. If this assumption produces any error in our estimates, we suspect it is to understate inequality, since teams near cities with higher incomes to begin with may well be more likely to sell to urban consumers and to have close kinsmen at work in town. Nevertheless, this possibility is not out of line with the conclusion that is to be drawn from this exercise anyway—that the Chinese countryside is relatively equal but not all that equal in comparison to some other developing societies.

The first estimate concerns per capita incomes. It uses for an estimate of local inequality the distribution of per capita total income from Wugong's team three. This distribution was reported in abbreviated form in the last column of table 8.1. As already suggested, Wugong's distributions are not that dissimilar to a number of other studies conducted throughout China. If anything, it may be toward the more unequal of the normal range of reported inequalities in a single team. But the distribution of per capita *total* family income is used here, including private-sector income and remittances from urban kin. Because this distribution is somewhat more equal than the distribution of collective income alone, the estimate is probably fairly representative of the situation prevailing throughout the Chinese countryside.

In the per capita estimate, deriving from the operations reported in the methodological note to this chapter, the richest 10 percent of all rural households get 25 percent of all income while the poorest 40 percent get 21 percent. The gini coefficient for this distribution is .31. How should these figures be assessed?

The raw figures might suggest greater inequality in 1979 than in 1952, when the data in table 8.4 were collected. But the two sets of figures cannot be directly compared. In preparing the figures in table 8.4, Roll had only very crude data on six regions to index spatial inequality, and local inequality was indexed only by differences among peasants by class origins. These sorts of crude categories tend to underestimate total inequality, giving lower estimates of inequality in 1952 than in 1979 even if actual inequality remained constant. Thus the reaction to the comparison between our 1979 estimate and that for 1952 may be either, ''Ho-hum, its all in the methods,'' or, ''Aha, we were right. The failure to control spatial inequality has exacerbated overall inequality.'' A little of both reactions is called for. The change in methods for 1952 and 1979 overstates the worsening situation. But the changing pattern of spatial inequality also caused overall rural inequality conditions either to remain constant or actually to get worse over time.

Comparisons with other countries provide an additional way of assessing China's rural inequality in 1979. For this comparison data on total household income must be used—with the "total" here meaning both the combined income of all household members and the combined income from collective, private, and remittance sources. In this estimate, the richest 10 percent of all households get 28 percent of all income, while the poorest 40 percent get 16 percent. The gini coefficient is .37 (table 8.5).

By this standard, the Chinese countryside appears to be only modestly equal in comparison to many other developing societies. It is clearly more equal than the second group of Asian societies in table 8.5, and also more equal than most of the non-Asian developing countries. But it

Table 8.5

**Rural Income Inequality in China and Other Countries**

| | Percent income earned by: | | | |
| | Poorest 40% | Richest 20% | Richest 10% | Gini coefficient |
|---|---|---|---|---|
| China, 1979 | 16 | 44 | 28 | 0.37 |
| Asian countries: | | | | |
| Taiwan, 1972 | 22 | 39 | 24 | 0.29 |
| Pakistan, 1970-71 | 22 | 39 | 24 | 0.30 |
| S. Korea, 1971 | 21 | 39 | 23 | 0.31 |
| Bangladesh, 1966-67 | 20 | 42 | 26 | 0.33 |
| Sri Lanka, 1969-70 | 19 | 42 | 26 | 0.35 |
| Philippines, 1971 | 17 | 47 | 32 | 0.39 |
| Indonesia, 1976 | 16 | 46 | 32 | 0.40 |
| Thailand, 1970 | 14 | 51 | 34 | 0.45 |
| Malaysia, 1970 | 13 | 52 | 36 | 0.48 |
| India, 1967 | 13 | 53 | 36 | 0.48 |
| Non-Asian countries: | | | | |
| Costa Rica, 1971 | 18 | 44 | 28 | 0.37 |
| Mexico, 1963 | 13 | 55 | 38 | 0.48 |
| Honduras, 1967 | 13 | 55 | 38 | 0.49 |

*Sources*: China—see appendix of this chapter. Other countries—Shail Jain, *Size Distribution of Income* (Washington, D.C.: International Bank for Reconstruction and Development, 1975).

*Note*: Figures are the distribution of households by total household income.

is less equal than most in the first group of Asian societies. In part, the first group has less inequality because these countries are geographically smaller and have less geographic variability in climate and terrain.[22] But this only further substantiates the argument that it is spatial inequality in China that inhibits overall equality.[23]

This comparison, it should be noted, is only for current income. It leaves out important relative welfare aspects of these different societies. These welfare aspects include not only health and education, which are far more widely provided in China's countryside than in most other developing societies at comparable income levels, but also the provision of a basic consumption floor below which families are not allowed to fall. The Chinese state had serious problems in providing this floor out of state resources in many rural areas in the early sixties, and again in some places in the seventies. But thus far in the eighties the state has stepped in more frequently to provide grain to grain-deficit areas and even, in the case of severe drought and flood, has turned for the first time to the United Nations and other disaster relief agencies for aid.[24] Another factor affecting rural inequality is that for many years poor households have been able to draw subsistence rations even when they lack current income to pay for them. In many instances, debts incurred to the team were never paid back.

By these sorts of subsistence guarantees and other public service measures, then, China has been more equal—or has done better by its rural poor—than the figures in table 8.5 would suggest. Nevertheless, in many ways the figures also attest to a significant reality, as frequent reports in the press indicate. Those in chronic poverty-stricken areas have suffered severely in poor diets, clothing, housing, and other necessities as well as in access to education, culture, and other amenities relative to those in more amply provided regions.

Table 8.6

**Distribution of Rural Households by Per Capita Income***

| Income groups (yuan) | 1978 | 1979 | 1980 | 1981 | 1982 | 1983 |
|---|---|---|---|---|---|---|
| 100 | 33.3% | 19.3% | 9.8% | 4.7% | 2.7% | 1.4% |
| 100 − | 31.7 | 24.2 | 24.7 | 14.9 | 8.1 | 6.2 |
| 150 − | 17.6 | 29.0 | 27.1 | 23.0 | 16.0 | 13.1 |
| 200 − | 15.0 | 20.4 | 25.3 | 34.8 | 37.0 | 32.9 |
| 300 − | | 5.0 | 8.6 | 14.4 | 20.8 | 22.9 |
| 400 − | 2.4 | 1.5 | 2.9 | 5.0 | 8.7 | 11.6 |
| 500 + | | 0.6 | 1.6 | 3.2 | 6.7 | 11.9 |
| | 100.0% | 100.0% | 100.0% | 100.0% | 100.0% | 100.0% |
| Average income | 134 | 160 | 191 | 223 | 270 | 310 |
| Gini coefficient | .28 | .26 | .25 | .23 | .22 | .22 |
| Sample size | 34,961 | 58,153 | 88,090 | 101,998 | 142,286 | 165,131 |

*Source*: Gini coefficients estimated from the income figures, which are in *Brilliant 35 Years* (Beijing: China Statistical Publishing House, 1984).

What has happened since 1979? A recent report on Chinese survey results gives a first impression that will be surprising to many people but which fits some of the arguments in this paper. With the increased emphasis on family farming, the gaps in income within any one village may well have increased—the reports of jealousy of rich members and the attempts to confiscate their resources in the initial years of the new agricultural policies supports this suspicion. Nevertheless, allowing poor villages to shift into commercial crops and other products more appropriate to their locale could have helped narrow the gap between rich and poor locales even while the rich were prospering themselves.

All villages did prosper between 1978 and 1983. Including earnings in off-farm activities, remittances from cities, and other miscellaneous sources of income, per capita rural income increased from an annual average of 134 yuan in the first year to 310 yuan in the last (see table 8.6).[25] There was a tremendous jump in prosperous peasants—the percentage of families with a per capita income exceeding 400 yuan jumped from less than one-half of a percent in 1978 to over 23 percent in 1984. But the poor profited as well—while one-third of all families had per capita incomes below 100 yuan in 1978, less than two percent were this poor in 1984. The net result, as measured by gini coefficients, was that overall income inequality declined steadily since 1978, from a high of .28 to a low of .22 in the most recent year.[26] With this increased equality China may well have moved toward the egalitarian country-side which many people long thought it to have. And these trends again suggest that in income distribution, things are not always what they seem.

The income distribution statistics do not tell the full story. There has also been a weakening of rural education, health care, and welfare benefits that provided a floor under living standards. The rapid increase in income differences within (as opposed to between) villages has stimulated conflicts in some villages. But the rapid increase in income and opportunities for most farmers may more than offset any feeling of injustice among the great majority of the rural population.

## Methodological Note

The estimate of combined rural inequality starts with a 1979 survey of collective income in almost all of China's five-million-plus rural pro-duction teams (table 8.6, columns a and b). This distribution is based on the internal accounting prices of grain distributed in kind to the members of each production team. In poor teams, which distribute little cash income, that distribution looms much larger than in rich teams. Because the internal distribution price is below the state pur-chase price of grain and other goods, rural income is overstated and particularly so in poor teams. To correct for this we apply an estimate of

Table 8.7

## Distribution of Per Capita Collective Income Among Rural Production Teams, 1979

| Income groups by average per capita income (yuan) | Percent of all teams | Percent income in kind | Adjusted per capita income | Households per team | Percentage of all households |
|---|---|---|---|---|---|
| (a) | (b) | (c) | (d) | (e) | (f) |
| 350 | 2.3 | 0.20 | 369 | 40 | 3 |
| 200 | 5.3 | 0.35 | 219 | 40 | 6 |
| 120 | 17.2 | 0.50 | 136 | 32 | 17 |
| 85 | 18.2 | 0.65 | 100 | 32 | 18 |
| 60 | 29.8 | 0.80 | 73 | 32 | 29 |
| 45 | 19.1 | 0.95 | 57 | 32 | 19 |
| 35 | 8.2 | 1.00 | 44 | 32 | 8 |
| 84 | 100.1 | | | | 100 |

*Sources and derivations by column:*
a, b) Ministry of Agriculture survey of almost all of China's 5 million plus rural production units as adapted in E. B. Vermeer, "Income Differentials in Rural China," *China Quarterly* 89 (1982):14.
c, e) Estimates.
d) a + (a * c * .27). An adjustment for the market value of income in kind.
f) (b * e)/(sum of b * e for the whole column).

percent of income distributed in kind (column c) and then derive an adjusted per capita income (column d) by the formula adusted income = per capita income + (percent in kind * per capita income * .27), where .27 is the proportional understatement in price and "*" is the multiplication sign. Another needed adjustment is for the size of a team, which appears to increase in rich areas that can support more population (column e). A rough estimate of team size (column e) multiplied by the percent distribution of teams by income group (column b) gives a new estimate of the number of households in each spatial income group (column f).

The next step in this estimation procedure is to enter columns d and f from table 8.6 into the first and last columns of table 8.7. The first column in table 8.7 provides the spatial dimension of the combined estimate. The last column has been divided by 10 to give the percentage of households in each of the ten interior cells. For example, in the first interior row, 3 percent of all teams earn an average income of 369

Table 8.8

## Merging of Local and Spatial Inequality to Derive Total Per Capita Collective Income Inequality

| Spatial inequality# | Local Inequality* | | | | | | | | | | Percentage households† |
|---|---|---|---|---|---|---|---|---|---|---|---|
| | .44 | .66 | .78 | .84 | .93 | 1.00 | 1.07 | 1.19 | 1.36 | 1.78 | |
| | | | | | | yuan per capita** | | | | | |
| 369 | 162.3 | 243.5 | 287.7 | 309.9 | 343.1 | 368.9 | 394.7 | 439.0 | 501.7 | 641.9 | 0.3 |
| 219 | 96.3 | 144.5 | 170.7 | 183.9 | 203.6 | 218.9 | 234.2 | 260.5 | 297.7 | 380.9 | 0.6 |
| 136 | 59.9 | 89.9 | 106.2 | 114.4 | 126.7 | 136.2 | 145.7 | 162.1 | 185.2 | 237.0 | 1.7 |
| 100 | 44.0 | 65.9 | 77.9 | 83.9 | 92.9 | 99.9 | 106.9 | 118.9 | 135.9 | 173.9 | 1.8 |
| 73 | 32.1 | 48.2 | 56.9 | 61.3 | 67.9 | 73.0 | 78.1 | 86.8 | 99.2 | 127.0 | 2.9 |
| 57 | 24.9 | 37.3 | 44.1 | 47.5 | 52.6 | 56.5 | 60.5 | 67.3 | 76.9 | 98.4 | 1.9 |
| 44 | 19.6 | 29.3 | 34.7 | 37.3 | 41.3 | 44.5 | 47.6 | 52.9 | 60.5 | 77.3 | 0.8 |
| Household members—>‡‡ | 4.5 | 4.6 | 4.9 | 5.0 | 4.8 | 4.3 | 4.1 | 4.0 | 4.0 | 2.8 | |

*Row below is per capita toal family income as ratio of the team average as determined in the Wugong survey. Families arranged from the poorest on left to richest on the right.

**Cells below derived by multiplying row above by average income in far left-hand column.

†Percentage of households in each cell to the left of the given figure. Derived from table 8.1 and from the fact that each column represents 10 percent of all households in a village.

‡‡Average household membershipp for each income group in Wugong survey, with some rounding for uneveness in household size.

#The left-hand column is average per capita income per team as reported in Table 8.1.

yuan. But these teams are divided internally among ten different income groups, each accounting for 0.3 percent of all households, and it is the income earned by each of these household groups that is shown in the interior of this table.

Before the per capita income earned by each of these groups is calculated, "local inequality" must be entered across the top of the table in the first row. The figures in this row are for collective per capita income in Wugong's team three as reported in table 8.1—only this time the figures are for decile (10 percent) income groups, arranged from poorest to richest 10 percent of all households and with their income expressed as a ratio relative to average team income. The top "local" row is then multiplied against the left "spatial" column to give the figures in each cell in the body of the table. For example, the adjusted per capita income of 369 yuan at the left multiplied by the ratio .44 at the top gives the upper left cell in the body of the table of 162.3 yuan per year for the poorest peasants within the richest group of production teams. The far right-hand column gives the additional information that this poorest-of the-richest group of households constitutes 0.3 percent of all peasant households.

Overall, the poorest families tend to be at the bottom left of the interior cells of this table and the richest at the top right part of the table. This ordering by per capita income provides a means of rearranging the families sequentially from poorest to richest. With this ordering, we construct two columns of figures, containing figures for seventy household groups arranged from poorest to richest—these columns are not shown here. The first column gives the total income earned by each household group. It is calculated by multiplying the per capita income of the group by its average household size (as given in the last row of table 8.7) and by the size of the group (as given in the last column). Smaller households have higher *per capita* incomes because they share income with fewer children and retired old people. The percentage of households shown in the right column is for each cell in its row—the percentage is constant among the cells in each row because each column represents precisely 10 percent of all village households.

The second column used in the final calculating step is simply the percentage of households in each group—that is, the figure from the right-hand column of table 8.7. This column, summed from top to bottom, gives the poorest 10 percent of all households, the poorest 20 percent of all households, and so on up to 100 percent of all households. Comparisons of this cumulative column with a similar summing and subsequent percentaging of the left calculating column gives the percentage of total income earned by each successive percentile of households. The gini coefficient is calculated from the same two columns.

These are the steps in computing the distribution of per capita incomes in the countryside. The steps in computing the distribution of total family incomes are similar, only that the first row of table 8.7 is for total family income and the last row for family membership has a rather different progression of figures in it. The internal cells are for total family income, and they are calculated by multiplying the left-hand column by the top row of ratios and by family membership.

## Notes

1. Dwight Perkins, "Meeting Basic Needs in the People's Republic of China," *World Development* 6 (1978):562. Perkins proceeds to note, however, that on a national basis rural income differentials have changed little since land reform.

2. Alexander Eckstein, "The Chinese Development Model," in Joint Economic Committee of Congress, *Chinese Economy Post-Mao* (Washington, D.C.: GPO, 1978), p. 102.

3. See particularly Simon Kuznets, "Quantitative Aspects of the Economic Growth of Nations: Distribution of Income by Size," *Economic Development and Cultural Change* 11, 2, part 2 (January 1963):1–80; Irma Adelman and Cynthia Taft Morris, *Economic Growth and Social Equity in Developing Countries* (Stanford: Stanford University Press, 1973); Albert Hirschman, *The Strategy of Economic Development* (New Haven: Yale University Press, 1958); Gunnar Myrdal, *Economic Theory and Underdeveloped Regions* (London: Duckworth, 1957).

4. The team three survey was conducted by Edward Friedman, Kay Johnson, Paul Pickowicz, and myself in preparation of our study, "Wugong: A Chinese Village in a Socialist State" (forthcoming). I am grateful to my collaborators for permission to use a portion of the data here. The team two data were graciously supplied by Peter Nolan from the results of a British survey team in 1978. The team two data are restricted to household collective income excluding both per capita incomes and incomes from extracollective sources.

5. There are methodological problems here that cannot be resolved utilizing data available at present. Our data take the household as the unit of income but conceal income pooling in larger networks of family, particularly financial support by sons for aging parents. However, the available comparative data likewise use the household unit in most cases. Our data also do not disclose payments by the team to "five-guarantee households" and other welfare recipients—a very important factor in assuring subsistence.

6. *Beijing Review* 25 (1982):7. Private-sector earnings rose rapidly in the late seventies and early eighties; Wugong data for 1981 would undoubtedly show a higher percentage of private-sector earnings.

7. Keith Griffin and Ashwani Saith, "The Pattern of Income Inequality in Rural China," *Oxford Economic Papers* 34, 1 (March 1982):175–79; E.B. Vermeer, "Income Differentials in Rural China," *China Quarterly* 89 (March 1982):20.

8. This finding is in line with the broad conclusions drawn by Keith Griffin and Ashwani Saith in *Growth and Inequality in Rural China* (Geneva: ILO, 1981), p. 16. We should be wary of generalization on this point, however, and not only because of the limited data available. Viewed in spatial terms, income earned in the private sector

probably exacerbates certain inequalities among units and among regions.

9. Peter Nolan and Gordon White, "Distribution and Development in China," *Bulletin of Concerned Asian Scholars* 13, 3 (1981):4.

10. John Gartrell, "Inequality Within Rural Communities of India," *American Sociological Review* 46 (1981):776.

11. Alfonso Castro, N. Thomas Hakansson, and David Brokensha, "Indicators of Rural Inequality," *World Development* 9, 5 (1981):402; Keith Griffin, *The Political Economy of Agrarian Change* (London: Macmillan, 1979), 2d ed. Louis Kriesberg notes the high correlation between income inequality and land inequality (.59) in the fifteen countries for which data existed for both indicators, *Social Inequality* (Englewood Cliffs, N.J.: Prentice Hall, 1979), p. 99.

12. Charles Roll, "Incentives and Motivation in China," paper presented at the annual meeting of the American Economic Association, December 28, 1975, p. 35, cited in Bruce Stone, "China's 1985 Foodgrain Production Target: Issues and Prospects," in *Food Production in the People's Republic of China*, ed. Anthony Tang and Bruce Stone (Washington, D.C.: International Food Policy Research, 1980), p. 108.

13. Li Chengrui, *Draft History of the Agricultural Tax in the People's Republic of China* (Beijing: Finance Publishing House, 1959), as adapted by Charles R. Roll, *The Distribution of Incomes in China* (New York: Garland, 1980), pp. 60–63, 75.

14. Cf. John Logan's illuminating discussion of "Growth, Politics and Stratification of Places," *American Journal of Sociology* 84 (1978):404–16.

15. Nicholas R. Lardy, "Regional Growth and Income Distribution in China," in *China's Development Experience in Comparative Perspective,* ed. Robert Dernberger (Cambridge: Harvard University Press, 1980), p. 167. See also Lardy's *Economic Growth and Distribution in China* (Cambridge: Cambridge University Press, 1978).

16. Lardy, "Regional Growth and Income Distribution in China," p. 171. Lardy's conclusions have been challenged by Audrey Donnithorne, "Centralization and Decentralization in China's Fiscal Management," *China Quarterly* 66 (June 1976):328–40.

17. Cf. E.B. Vermeer's insightful discussion of a number of the issues touched on here in "Rural Economic Change and the Role of the State in China, 1962–78," *Asian Survey* 22, 9 (September 1982):823–42. Robert Hsu, "Agricultural Financial Policies in China, 1949–1980," *Asian Survey* 22, 7 (July 1982):638–58.

18. Suzanne Paine, "Spatial Aspects of Chinese Development: Issues, Outcomes and Politics 1949–79," *Journal of Development Studies* 17 (1981):138.

19. The changes in rural organization in the early 1980s create new conditions which are beyond the scope of the present chapter.

20. This may help to account for the somewhat surprising fact that in the early eighties the strongest pressures from below to dismantle collective agriculture in favor of household-based economy came from poorer, frequently mountain, communities.

21. *Xinhua yuebao*, February 1981, pp. 117–20. Cf. the discussion in Vermeer, "Income Differentials in Rural China," pp. 25–29.

22. There may be additional methodological problems in how the data were collected in each country—e.g., I remain suspicious of the Bangladesh and Pakistan figures. Unfortunately, too little detail is given in the Jain source to deal with this suspicion.

23. These comparisons are less favorable to China than those in the World Bank study, *China: Socialist Economic Development* (Washington, D.C.: World Bank, 1983). However, the World Bank study compares China's *per capita* rural income distribution with the *total household* income distribution of several other societies, thereby overstating China's rural equality. And the spatial inequality figures used in the World Bank estimate for China are an earlier, less complete set that also overstate China's rural equality.

24. United Nations, Office of the Disaster Relief Organization, *China, Case Report: Drought and Floods in the People's Republic of China* (n.p., 1981); *China, Case*

*Report: Drought and Floods in Hebei/Hubei Provinces 1980/81,* Case Report no. 11, May 1982.

25. Data from *Brilliant 35 Years* (Beijing: China Statistical Publishing House, 1984), as cited by Fan Kang, University of Chicago Agricultural Economics Workshop, November 1984. There are two potential concerns about these survey data. The more serious is that there is about one extra person per household in these data compared to the Census and other data sources. This could be a legitimate difference, but it could also suggest that the surveys oversampled prosperous, able-bodied, intact households and left aside the elderly and other poor single-family homes. Another potential concern is that the sample expanded rapidly over these years, perhaps maintaining the same sampling frame but also possibly incorporating new locales that were more equal than those in the early surveys.

26. These gini coefficients are estimated from the original Chinese income data. The high figure of .28 for 1978 or .26 for 1979 is lower than our earlier estimate of .31 for 1979 per capita income, suggesting that we may have overestimated inequality somewhat. However, the discrepancy is so small as not to change our original conclusions.

# 9

## Balance and Cleavage in Urban-Rural Relations

## Marc Blecher

Embedded in the theoretical literature are two very different ways of thinking about the relationship between city and countryside. They are usually conflated, which leads to analytical confusion, promotes needless disputation, and obstructs a more sophisticated understanding of the *problematique* in general. On the one hand there are questions of *balance*: How do rural and urban incomes and standards of living stack up? What are the financial flows between them? How are the economic, human, and social and cultural costs of industrialization distributed?

Earlier drafts of this chapter were presented at the Conference on Bureaucracy and Chinese Rural Development, University of Chicago Center for Far Eastern Studies, August 26-30, 1981, and at the Conference on China in Crisis, Contemporary China Center, Oxford University, September 7-10, 1982. It is based upon collaborative research carried out by Phyllis Andors, Stephen Andors, Mitch Meisner, Vivienne Shue, and me in China in 1979. Our field trip was arranged at the invitation of the Chinese People's Association for Friendship with Foreign Countries, cosponsored by the Chinese Academy of Social Sciences, and funded by generous grants from the National Endowment for the Humanities and the Ford Foundation. Oberlin College also helped finance my own participation in this project with a generous Powers Travel Grant. We wish to thank the officials and citizens of Shulu County for their extraordinary cooperation during our stay there. This chapter was drafted by me alone, but it draws upon material contained in chapters of *Town and Country in a Developing Chinese County*, a forthcoming collaborative book on our trip, which were drafted by other members of the group and me. The interpretations and analyses in this chapter as well as any errors are my responsibility alone. I owe a debt of gratitude to the participants in both conferences, who provided stimulating discussion regarding some of the issues raised here and many others relating to them, and in particular to Bill Parish and Shigeru Ishikawa for their detailed, careful, and thoughtful comments. The information upon which this chapter is based was current as of 1979. Subsequent follow-up research has not been possible, so the extent of change is not known. This should be kept in mind in reading this chapter, in which the present tense is used to describe findings whose currency may sometimes be in question.

What are the distributive or redistributive effects of state policy and action? On the other hand are questions of *structural cleavage*: What separates rural and urban people, at all levels of the social structure (landlords vs. capitalists, middle peasants vs. the petit bourgeoisie, poor peasants or rural proletarians vs. urban workers)? What different opportunity structures and mobility prospects face them? What are the different ways in which they are organized within their respective modes of production and within the political arena?

This distinction between urban-rural balance and cleavage reveals much about the Chinese case. By all accounts, Chinese socialist economic development is generally regarded as having been fairly successful in promoting urban-rural balance. Looking at the entire period from 1949 to 1976, Suzanne Paine finds no evidence of systematic urban bias in the sense of planning decisions and resource allocations that favored urban areas at the expense of causing suboptimal results for the development of the economy as a whole.[1] William Parish and Martin Whyte speak of real progress in reducing the urban-rural differential in income, orientation of production, education, and public health.[2] The urban-rural income gap in China is around 2.6:1 or 2.7:1, which is low in comparison with other developing countries.[3] Urban-rural terms of trade, though still working against the rural areas (at least according to no less authoritative a Chinese economic planner than Xue Muqiao),[4] have nevertheless moved consistently in favor of the rural areas since 1949.[5] Rhoades Murphey has summed it up: "The Chinese have been more successful than the Indians, or than most other countries . . . in distributing the benefits of growth more adequately to all regions of the country and to the rural areas."[6] So while of course much progress remains to be made, China has already chalked up significant achievements in the area of urban-rural balance.

But the Chinese case looks rather different in terms of the structure of the urban-rural cleavage. First, different forms of political and economic organization have prevailed since the earliest post-Liberation days: in towns and cities the state form of production has predominated, supplemented by urban "collective" forms, some under close state supervision and regulation; in the countryside there have been rural collectives with certain (though of course variable) degrees of juridical, administrative, economic, and political autonomy, supplemented by elements of household economy. The post-Mao reforms have opened the cleavage still further, since in the countryside they have undercut collective forms of production and given new strength to the household economy but have been less successful in reforming the urban state economy. Second, earlier attempts to attack directly the social and cultural cleavage between town and country—such as the May 7 cadre schools, rustication of urban youth, and restructuring the organization of medical care and education—met with mixed results and have now been largely abandoned. Third, the classification of the Chinese people into administrative categories of rural and nonrural householders (*nongye hukou* and *fei nongye hukou*)—perhaps the most vivid manifestation of the urban-rural cleavage in China—remains intact and as difficult to cross as ever. It continues to determine into which of the two major constellations of residence, employment, social services, mode of production, and political organization—rural or urban—a Chinese citizen will be incorporated.

The general picture that emerges for China as a whole, then, is of a country that is comparatively egalitarian in terms of the economic balance between urban and rural, but at the same time characterized by a clear and, for the vast majority of the citizenry, largely insurmountable urban-rural cleavage. But of course in China what appears true at the aggregate or national level may appear less so at the middle or local levels. This chapter addresses the problem of urban-rural balance and cleavage in one rather ordinary Chinese county.

Shulu *xian* is a rural county in what was formerly easternmost Shijiazhuang prefecture, 60 kilometers east of Shijiazhuang city, in south-central Hebei province. In 1978, it was about average for the area in its agricultural base but somewhat above average in industrial development. Gross value of industrial output (GVIO) was 148,520,000 yuan, the highest in the prefecture;[7] per capita GVIO was 302 yuan. Gross value of agricultural output (GVAO) was 118,026,000 yuan, or 240 yuan per capita. The ratio of GVIO to GVAO was the second highest in the prefecture. But, for reasons discussed elsewhere,[8] per capita collectively distributed income in the rural collective sector was only 88.20 yuan. Shulu appeared in 1978, then, as a county with a satisfactory level of agricultural development and a very good one in industry.

Before proceeding, a note on definitions of urban and rural may be in order. Since the English language contains no word that captures a sense of places that fall between "urban" and "rural," they must inevitably be described as oxymoronic variants of one or the other: "large towns" or "small cities." There is no use in entering here into the largely sterile, idealist debate over what really constitutes a city and what doesn't. Still, the question cannot be avoided altogether by adopting the language and concepts of central place hierarchies; though of great value in posing many sorts of problems, they are of less use in raising others such as the relations and cleavages between peasants and workers, or between rural collective and urban state or collective modes of production and political formations. To face but not belabor the definitional question, some of the basic conventions of the literature on urban-rural questions are followed here. Accordingly, places with populations over about 10,000, a very small proportion of whom work in agriculture or occupations directly linked to it, with a certain compactness and density of population settlement on the land, and with several central functions (like marketing, administration, social life, transport, or production) are classified as urban. By these criteria, Shulu's capital Xinji is clearly an urban area—we may call it a small city. Settlements that are clearly not villages (small, nested settlements of agriculturalists with perhaps a few people engaged in agro-related occupations) but also do not fall unambiguously into the urban category because of their smaller size, more heavily agrarian economic base, or lower centrality are in the intermediate zone between urban and rural— we may call them towns. Shulu has quite a few of these. They are not strictly "urban," but they do have aspects or features of urbanity, just as they do of rurality. It is possible, then, to speak of Shulu county as a place with at least one city, some semi-urban, semi-rural towns, and many villages. It is an area where urban and rural mix and meet.

## Urban and Rural in Shulu County

Shulu county contains one main urban center and twelve other medium-sized towns larger than ordinary commune seats. The major center is Xinji town, at once a center of political and state administration for the county as a whole, a locus of subcounty services for five neighboring communes, a commune headquarters, the location of the main Shulu commercial outlets, an already large and still rapidly growing industrial center. It is a major station on the Shijiazhuang-Jinan railway line. In 1978 it had an official population of 27,269, but its actual population was significantly greater than this. It also included a large though indeterminate percentage of the 9,994 contract workers employed in industry and commerce who resided in Xinji most of the time, although their families and permanent homes were in the countryside. In other words, around 35,000 people lived in Xinji proper in 1978. Another 16,686 people—the residents of Chengguan Commune—lived in the environs of Xinji, not formally part of the city but, at least in the areas around the city limits, physically indistinguishable from it. So, what could be called greater Xinji—i.e., the built-up urban area within and around the city boundaries (not including its suburban settlements and farms)—had around 50,000 people in 1978.

Downtown Xinji includes government offices, industrial plants, a number of large department stores, smaller specialized shops, restaurants and tea houses, a large new three-story government guest house *cum* meeting center, a public library, a hotel (with a flashing electrical sign proclaiming its name—''The Victory''), a bus station, and a new, capacious, aircooled movie theatre; a hospital lies at the edge of town. A free market operates daily, with a larger periodic market fair every five days. Xinji has the look and feel of an industrial boom town. Tall smokestacks dot its skyline, spewing emissions that elicit pride among the people of an industrializing culture (and apprehension among visitors from a postindustrializing one); bustling road traffic comprises a marvelous array of vehicles, including oxcarts, bicycles, hand tractors pulling trailers, rickety buses, and air-conditioned Japanese vans with foreign guests peering with intense curiosity out the tinted glass windows. Streets are made messy either by the latest rain or by construction of new sewer lines, or both. One-story buildings which look like they date from the Tang (but are probably from the early socialist period) are punctuated by multistory, modern but functionally plain structures. Xinji shows the clear architectural and visual signs of a place that has been growing very quickly of late and is still doing so.

As of mid-1979, Xinji People's Commune stood in a very interesting relationship to the city in which it is located. In particular, the relationship demonstrates the flexibility with which the differences that are supposed to inhere in the formal distinction between rural and nonrural household registration were applied in practice. Let us turn briefly and parenthetically to these formal differences. The category of rural residents included those who are members of people's communes by virtue of having been included in the communes upon their establishment or having been born into rural households subsequently. These people were employed in their rural collective units, and legally they could work outside of them only under a contract by that unit with an outside employer. Social services such as education, medical care, and welfare were also provided them through their rural collective units. They could not take up permanent residence outside their commune, although those with contract employment in the towns and cities often resided at their places of work on a temporary basis. Perhaps most important, they derived their supplies of grain and other rationed necessities (like oil and cloth) from their collective units, in the case of grain at a low price internal to the collective unit below the state retail price. Their political life, including political study, mediation of disputes, participation in mass movements, and civil administration, was handled through the rural collective units. The category of nonrural residents included almost everyone else (except people in special categories such as returned overseas Chinese and foreigners). Usually, such people were employed in state-run or collective units administered by government at the county level or above, or in commune (and now *xiang*) governments. They received social services either from the units that employed them or from the government of their town or city. Their political lives also revolved around these units. They purchased their supplies of grain and other rationed necessities from government-run outlets in the towns or cities.

Xinji People's Commune's per-hectare yields of grain and cotton were the highest of any commune in the county. But despite its rural (people's commune) organizational structure and the classification of its members as rural residents, it was in many ways more an urban, industrial than a rural, agricultural unit. It contained only .04 hectares of cultivated land per capita, the lowest in the county. While during busy seasons 98 percent of its labor force worked in agriculture, during other times of the year 80 percent worked in one or another enterprise, some of which were related to agriculture (like food processing or husbandry of pigs, dairy cows, and even mink), but others of which were not (such as the production of penicillin ampules, glue tubes, and bicycle axle sheathings). The commune was divided into ten brigades, each named after one of a numerical series of urban streets—First Street Brigade, Second Street Brigade, etc. The residential neighborhoods of the commune members (rural residents) were interspersed with the city (nonrural) residents', and in some cases even their houses were interspersed in the same neighborhood. Some nonrural residents rented their housing from the commune. It is especially notable that in this urban center there were no urban residents' or neighborhood committees as such. Political work on a neighborhood level among nonrural residents of Xinji town was carried out by the brigades and teams of Xinji People's Commune, and all interpersonal disputes, including those between rural and nonrural residents, were mediated by civil adjustment committees (*minshi tiaojie weiyuanhui*) under the brigades. Nonrural residents could seek medical care from the commune clinic or hospital. The children of rural and nonrural residents attended school together. In most ways, then, Xinji Commune was relatively urbanized, and in fact remarkably well integrated with the administrative and social service apparatus of the urban government, actually performing many of the functions of that government. Moveover, the formal distinction between rural and nonrural residence, which is a cornerstone of Chinese policy controlling urban growth and a basic form of social stratification, was quite blurred in practice.

Shulu county also contains nearly thirty other settlements which fall into the intermediate zone between cities and villages. Before describing them it may be useful to review some of the analytical categories and themes of central place theory as it has been applied to the Chinese case.[9]

Rural China has historically possessed a complex marketing structure with a clear, layered hierarchy of levels of market towns. The lowest level of this structure (above the nucleated village [cun], which had no market, and the minor or "green vegetable" market [yaodian]) was what Skinner has called the standard market town, which he defines as "that type rural market which met all the normal trade needs of the peasant household."[10] It provided for horizontal trade of ordinary goods within the rural area that depended on it, and it was also the point at which commodities made their first entry into or their final exit from the larger marketing system that lay outside the local "standard marketing area." Above it are "intermediate" and "central market towns" (as well as large city and regional towns and market complexes beyond). The central market town is "normally situated at a strategic site in the transportation network and has important wholesaling functions," while the intermediate market aggregates trade between the central market above and the standard market below.[11] The overlay of the marketing with the administrative structure was imperfect at best; during the late Qing period (approximately the nineteenth century), county seats, which were the lowest levels of state administration, often coincided with the loci of central markets, but sometimes with intermediate markets. In the postimperial period, state administration was extended downward below the level of the county seat, in a pattern of dizzying complexity, change, and variation that need not concern us here. Suffice it to say that by 1957, the "township" (xiang) was consolidated as an important level of governance below the county, though its boundaries were subject to some further enlargement (by an average of about one-fourth in terms of population) by 1958. The objective of this particular phase of boundary-drawing was to bring the administrative structure into line with the marketing structure: township governments were to be located in market towns and their boundaries were to coincide with the rural areas served by those markets.

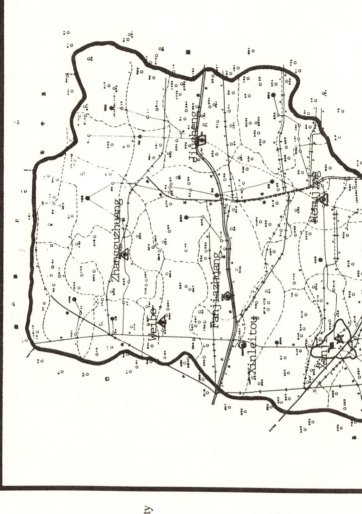

★  County seat

□  Town (*xiao chengzhen*) and county government administrative subcenter

△  Standard or intermediate market town (*zhen*) and county government administrative subcenter

○  Ordinary standard or intermediate market town (*zhen*) only

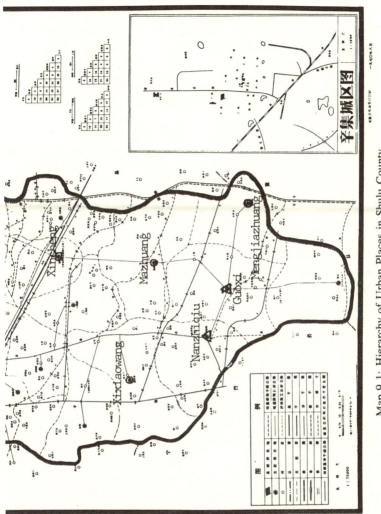

Map 9.1: Hierarchy of Urban Places in Shulu County

When the people's communes were first established in 1958, they generally coincided with the boundaries of the townships that had immediately preceded them; they also took over the townships' political and administrative functions and merged them with the economic functions (formerly handled by agricultural producer cooperatives) of organizing and managing production and distributing income to the peasants.

The two most important intermediate settlements in Shulu county were both former capitals of the county, named, appropriately enough, Jiucheng (Old City) and Xincheng (New City). They are referred today as *xiao chengzhen*, which translates literally as "small market city"; we may call them towns. Though neither has been a county seat since liberation, both have been headquarters of the former townships (*xiang*) as well as of the people's communes in their initial large and subsequent smaller incarnations. Both have a distinct articulation of criss-crossing downtown streets rather than just the single main drag characteristic of smaller settlements. They are marked with squares on map 9.1.

In addition, there were by the 1950s ten standard and intermediate market towns, of which five were headquarters of township governments and, later, of the early, large communes (marked with a △ on map 9.1). It appears, then, that while some of the smaller marketing areas and towns fell below the township (and early commune) level, all the townships (and early communes) were headquartered in market towns and probably coincided with their respective marketing areas. They leave a legacy to the present day. The five towns that were both marketing and administrative centers, together with Jiucheng and Xincheng, comprise a group of seven subcenters of county government administration and social services. Each of these has at least a post office, a bank branch office, a grain station, branch offices of the county Finance and Tax Bureau and the Industry and Commerce Management Bureau, a police substation, a courthouse, and a cinema. In 1979, these seven towns did not of course constitute an intermediate level of government between the county and the commune. They were simply sites of the extension outward into the countryside of county government administration. This is significant enough in itself, since it meant that county government in Shulu was not confined to the county seat, as has been standard in Chinese historical administrative and governmental practice even through the People's Republic period. But with the current political reforms involving the hiving off of political authority from the communes and their vesting in township governments, these seven towns may well be slated to become formal governmental centers in the near future.

These towns have other trappings of urbanity. While all thirty-one Shulu communes have hospitals, evidence from Jiucheng, whose hospital attracts referrals from three other Shulu communes (and even treats patients from neighboring Shen county to the east in Hengshui prefecture), suggests that some of the hospitals in these rural towns may be better staffed and equipped than the average commune hospital. Each of the seven towns had a small contingent of residents who are members of the county government staff (classified as nonrural residents) rather than the local people's commune: in Jiucheng there were 300 such persons. Some have county-run enterprises: Jiucheng, for example, has eighteen, including a cotton-oil extracting plant and an agricultural machinery plant, the members of whose labor forces are classified as nonrural residents.

It has often been remarked that China as a whole presents a pattern of urbanization rather different from many third world countries. Instead of having primate cities and then relatively few medium-sized ones between them and the small village settlements of the countryside, China possesses a more fully articulated pattern of large, medium, and small cities. Shulu seems to replicate this pattern on a smaller scale. Within the confines of this county, which is located well into the countryside, there is not just one urban center and a large rural periphery, but rather a complex hierarchy of urban places articulated in an overlapping structure of governmental, economic, social, and historical characteristics. Aside from Xinji, there are two larger towns and ten primary or secondary market towns, five of which, together with the two larger towns, are historical as well as contemporary centers of subcounty administration. These thirteen places as well as sixteen others also have more localized administrative, political, economic, and social functions as the seats of people's communes. In short, there are quite a number of settlements—some obviously more "urban" than others—dotting the Shulu countryside, bringing a variegated range of services, activities, and forms of productive, commercial, social, and political life normally associated with some degree of urbanity within reach of Shulu's predominantly rural and agricultural population. Clearly, they serve to narrow the distance—in the most literal sense—between urban and rural areas.

Yet there is a danger, particularly in the present climate of cost-conscious economic reform, that the towns at the lower end of this central-place hierarchy may tend to usurp the role of rural institutions in providing social services, economic initiatives, and political organizations and channels of expression. Indeed, this has already begun to happen in parts of China. On grounds of economic "rationality," commune hospitals, clinics, and schools are being closed down, and their resources and functions redirected to health and educational institutions in small towns. The same sort of thing is happening to rural enterprises and even governments. Of course, the variegated pattern of articulated central places at the local levels does not *cause* any of this, but it *provides the material basis* that makes it possible in the first place. So even as the existence and growth of small towns and urban places bring on the rural and urban areas into closer proximity, they may also serve to differentiate rural and urban life and institutions in new ways. In other words, small-scale urbanization may simultaneously narrow the urban-rural gap while also deepening the urban-rural cleavage.

## Urban-Rural Financial Flows Mediated Through State Agencies

Explication and calculation of the financial flows between urban and rural areas is a major issue in the analysis of urban-rural balance. It proves to be a very complex area of study, since it must include questions of trade (relating to composition, volume, and terms), private and public investment, private transfers (e.g., remissions home from emigrant peasants), and public transfers (e.g., differential taxation and expenditure, subsidies, services). Here no attempt is made to address either the general question of urban-rural financial balance or the more specific one of the urban-rural balance in government financial activities, which would, in a socialist country like China, have to include the effects of governmentally determined trade terms and flows and the whole range of state policies concerning wages, hidden and overt subsidies to various industries and sectors, and so forth. In this section our sights are more limited: we look at the urban-rural financial balance in the various *budgetary* processes in which Shulu county government agencies are bound up.

Even this matter is far from simple or even straightforward. There is no single budget or budgetary process through which all financial administration takes place in Shulu (or, for that matter, any Chinese county). The various channels and forms of financial flow can be grouped into three categories: the official state budget of the county government, the vertical systems (*xitong*) of production and administration, and the county's extrabudgetary (*yusuan wai*) operations and funds. Full description and analysis requires far more detail than is appropriate here.[12] This section covers only the barest outlines, and then proceeds to an analysis of the various financial flows in terms of the urban-rural question.

### Official State Budget of Shulu County Government

The state budget of the county government is certainly not the only or even the master budget affecting state agencies in Shulu; it may not be the most important one. It covers only certain very specific areas of revenue—all taxes but only certain enterprise profits—and of public expenditure.[13] It does not include the investment funds that are channeled into the county through the various functional bureaus engaged in overseeing the economy, such as the industry or commerce bureaus. It does not include revenues and expenditures of the vertical systems (*xitong*), discussed below. It does not include extrabudgetary funds either, also discussed below. Nevertheless, it is a very important document, summarizing a process of financial egress from and ingress to the county which involved a total of 36,173,000 yuan in 1978.

The amounts of expenditure and revenue in this budget are determined by higher levels of the state, with penalties and rewards to the county government depending on its successes in meeting them. Ever since 1949, Shulu county's official state budget has been in considerable surplus; in 1978, revenues exceeded expenditures by a factor of over three. In other words, as far as the specific areas covered by the official state budget goes, Shulu is a surplus-producing county, turning over to the higher level of state administration far more than it receives from the state.

But if the question of the relative balance (or in this case lack of balance) between state budgetary revenues and expenditures is posed for just the rural areas of Shulu, a rather different picture emerges. The rural collective sector provides two categories of revenue to the state. It pays all of the agricultural tax, which in 1978 amounted to 1,759,000 yuan, a mere 6.4 percent of total county revenues in the state budget that year. It also pays an income tax (*suode shui*) on a portion of the profits of brigade- and commune-run enterprises. This income tax is also paid by thirty of the forty-eight county-run industrial enterprises, though at over three times the rural rate. Commune and brigade enterprises paid about 47 percent of the total income tax revenues in 1978, amounting to about 2,059,000 yuan. Adding the agricultural tax, a total of about 3,820,000 yuan was contributed to the state budget. (It should be remembered that some of this comes from enterprises in Xinji Commune and other somewhat urbanized places, especially commune towns, in Shulu; the total rural contribution therefore is somewhat less.)

On the expenditure side of the state budget, there are four main categories. In order of their magnitude in 1978, they were culture, education, and health; support for agriculture; county government administration; and emergency welfare. Support for agriculture is unambiguously directed at the rural areas (with possible minor exception of any assistance received by Xinji Commune's small agricultural sector); in 1978 it totaled 2,584,000 yuan. In this category alone, then, the rural areas received back through the official state budget two-thirds of the monies they paid in. If we add in some fraction of the 2,928,000 yuan in expenditures on culture, education, and health and the 1,160,000 yuan for emergency welfare, the rural areas will be close to a balance—perhaps above, perhaps below—in financial flow in the official state budget of the county government. We have no way of estimating this fraction; for argument's sake, if we use a conservative figure of 50 percent of the total county expenditures on culture, education, health, and emergency welfare, the rural areas would strike a ratio of receipts to remittances of 1.2:1. Even if the actual ratio were somewhat lower than this, the rural area would nevertheless be well ahead of the urban areas of the county in terms of the official state budget, which, it will be recalled, registered a ratio of receipts to remittances for the entire county—urban and rural—of about .33:1.[14]

*A Vertical Administrative System: The People's Bank*

A second category of financial flow affecting Shulu county is the vertical systems (*xitong*). In contrast with the state budget, which collects revenues in a variety of categories and disburses them across a variety, keeping a large share for use at higher levels and moving funds from sector to sector, the vertical systems are unitary: each collects funds from a given set of units or enterprises under its jurisdiction and disburses them to that same set. Examples encountered in Shulu were the Industry and Commerce Bureau, the Second Light Industry Bureau, and the People's Bank. Of these, the only one that directly involves the urban-rural question is the People's Bank.

The urban-rural balance in the activities of the People's Bank is complex, mixed, and difficult to interpret. The terms attached to various categories of loans to agriculture and industry favor the former:

Table 9.1

### Shulu County Major People's Bank Loans, 1965, 1971, 1978 (unit: yuan)

| Type of Loan | 1965 | 1971 | 1978 |
|---|---|---|---|
| To state-run industries | 426,000 | 4,509,000 | 9,968,000 |
| To county-run collective industries | 39,000 | 1,251,000 | 700,000 |
| Cash flow loans to enterprises | 4,000 | — | 687,000 |
| To commercial enterprises | 5,002,000 | 4,580,000 | 8,701,000 |
| (Subtotal of loans to urban areas, with percentages of total lending) | (5,471,000 [32%]) | (10,340,000 [78%]) | (20,056,000 [75%]) |
| Advances to teams and brigades against coming harvest | 2,772,000 | 897,000 | 2,005,000 |
| To communes and brigades | 7,438,000 | 1,588,000 | 3,898,000 |
| To commune- and brigade-run enterprises | 733,000 | 28,000 | 30,000 |
| To credit cooperatives | 256,000 | 16,000 | 29,000 |
| To commune members | 462,000 | 320,000 | 849,000 |
| (Subtotal of loans to rural areas, with percentages of total lending) | (11,661,000 [68%]) | (2,849,000 [22%]) | (6,811,000 [25%]) |
| Total | 17,132,000 | 13,189,000 | 26,867,000 |

Table 9.2

### Shulu Savings Deposits (unit: yuan)

| | 1953-54 | 1979 | 1979 per capita (approx.) |
|---|---|---|---|
| Individual commune members | 2,002,000 | 7,950,000 | 16.87 |
| Commune, brigade, and team | 352,000 | 21,921,000 | — |
| Urban residents | — | 5,659,000 | 269.48 |
| Total | | 35,530,000 | |

agricultural loans had repayment schedules of up to fifteen years and interest rates between zero and .36 percent, while those to industry had maximum repayment schedules of two years and interest rates of .42 percent. Of course, the urban industrial sector has received much free capital through the economic ministries. Table 9.1 shows a decidedly urban tilt in the volume of loans extended[15] in both 1971 and 1978; the sharp departure from the rural tilt in the 1965 pattern reflects the pace and effects of rapid growth of Shulu's industrial and commercial sectors. In view of the fact that savings deposits in Shulu came overwhelmingly from the rural sector (at least in 1979—see table 9.2),[16] the imbalance seems even greater. Unknown, of course, are the urban and rural sectors' relative needs and demands for and capacities to absorb credit, which would have to be taken into account in interpreting these figures.

*Extrabudgetary Funds*

The third form of financial flow in Shulu is the county's extrabudgetary fund. Its sources are the after-tax profits of twelve county-run collective enterprises. These are enterprises that were set up in 1970 in response to the national policy calling upon localities to establish "five small" (*wu xiao*) industries, and they are not to be confused with enterprises owned and run by the rural collective sector, i.e., the communes and brigades. Since these enterprises pay the income tax, whose highest rate is 55 percent of profits, the monies available for the extrabudgetary fund can be quite large—in 1978, they totaled 2,928,000 yuan, which comes to one-fourth of total state budgetary expenditures in that year. These funds are available for expenditure by the Shulu County People's Government in any way it sees fit.

We were able to gather little detailed information on precisely how these funds are spent. County officials gave us the impression that much was spent on reinvestment for expansion of the twelve enterprises that had generated the funds. But this is not always the case. In 1978, for example, 2.5 million yuan were allotted from the extrabudgetary fund to finance half of the cost of a massive rural water conservancy project built in the county's historically drought-plagued southern region in the winter of 1977–78. This amounted to over 300,000 yuan more than extrabudgetary funds from year to year. It also gives us a feel for the large proportions of this urban-to-rural fiscal transfer: the water conservancy project soaked up more than one full year's net profits generated by twelve urban-based industrial and commercial enterprises.

*Budgetary Processes and Urban-Rural Balance*

Shulu has at least two different budgetary processes taking money out of the county and bringing money into it, and one (the extrabudgetary fund) moving money within it. Each includes rather different things and appears to strike rather different urban-rural balances.

It may bear reiterating at this point, though, that these budgetary processes do not comprise the sum total of state-mediated (not to mention private) financial flows in Shulu. If they did, we could simply total them up and strike the elusive bottom-line figure. In addition to these budgetary matters, state policies and actions affect the urban-rural financial balance in many ways. In general, the state still underprices agricultural products and overprices industrial ones. It gives differential subsidies to various industries in the rural and urban areas—some directly, some indirectly (by continuing to carry loss-making firms, by providing technical assistance and infrastructure, etc.). It supplies urban workers, not peasants, with pensions and fixed annual incomes. Precisely how each of these state activities affects urban-rural balance in Shulu is impossible to say without much further research.

## Temporary Contract Work: Rural Labor for the Urban Economy

One of the major issues in urban-rural relations is the flow of labor and population from the countryside to the cities in search of employment. China has attempted to limit this flow by instituting the household registration system described above. It has also sought to regulate the movement of rural labor into urban employment through a system of temporary contract work (*linshi hetong gong*), under which rural residents are recruited as workers for specific though often extended, continuing periods of time by urban industrial or commercial enterprises. These people retain their formal status as rural residents. They maintain their family residence in their teams, either commuting daily to work in the town or bunking during work periods in a factory dormitory and returning home on weekly or monthly visits.

Compared with the regular urban workers, they are subaltern in several ways. First, they are paid less than regular workers, averaging 455 yuan in 1978 compared with the latter's 571 yuan. (Still, the 455 was over twice the average peasant's annual income of 220 yuan.) Second, they do not even receive all of their lower wages. Before 1972, they were paid half of their wages in cash, and the other half went to their production teams, which in turn credited them with a full day's workpoints for each day worked. But after that year, a new system went into effect under which they were paid all of their wages in cash except for 3 or 4 yuan each month, which went to their teams. Both payment systems put them at an even greater economic disadvantage compared with regular workers than the mere wage differential suggests, though the post-1972 system ameliorated this somewhat. Third, they receive none of the fringe benefits that regular workers do, such as medical care or pensions. Fourth, they have less job security than regular workers, at least formally; while the latter have (at least up to now) had an "iron rice bowl"—virtual assurance against being fired—contract workers have held their jobs only under the terms of a contract which may or may not be renewed upon its expiry.

In fact, though, as in many places in China most "temporary contract work" in Shulu is not particularly "temporary"; contracts are usually renewed, and the ranks of contract workers have tended to expand rather than contract, at least in Shulu, since the system's inception there in 1964. In 1979, there was an impressively large complement of contract workers in Shulu. Numbering 7,779, they comprised a full 52 percent of the labor force in Shulu's large and growing state industrial sector. Another 2,215 made up 46 percent of the labor force employed in commercial work at the commune and county levels. Clearly, contract labor has played a very important role in providing the labor force for Shulu's impressive industrial and commercial growth.

The contract labor system has several advantages for the urban areas. First, it provides significant savings for urban employers, since contract workers are paid less than regular urban workers. In 1978 savings in urban industry came to around 900,000 yuan, or 12 percent of the total payroll. Second, since contract workers can be fired easily, urban enterprise managers gain a degree of flexibility and control over their labor forces which they are denied by the lifetime tenure of their regular urban workers. Third, it transfers much of the social reproduction cost of the urban labor force onto the rural areas. Contract workers can be housed relatively inexpensively in dormitories, usually on factory grounds, so that the supply of urban housing, which is already very scarce even for the existing complement of urban residents, need not be placed under greater stress. Schools do not have to be built in the urban areas to educate the children of contract workers, who stay home and attend rural schools. Contract workers have relied for their health care on the rural cooperative health plans, and aside from perhaps a minor expansion in the capacity of factory first-aid stations, no new medical facilities have to be provided in the city to treat contract workers. The employing units do not have to provide services for them as retirees. In short, the urban sector of both production units and city governmental agencies realizes considerable savings from this system, the magnitude of which can be apprehended from the fact that if the 9,994 Shulu contract workers all lived in Xinji with three dependents each, the population of the city proper would be around 250 percent larger. In view of the juridical and economic obstacles to expansion of the city's boundaries, it is a certainty that Xinji would under these conditions have become a very crowded and unpleasant place.

Of course, in the areas of wages, job security, and responsibility for costs of social reproduction of labor, the rural people and areas lose out with respect to their urban compatriots. But there are also rural benefits deriving from contract labor. First of all there are the economic transfers from urban to rural effected by contract labor. These take two specific forms: the monies appropriated by the rural collective units from the contract workers' wages, and the remittances sent back home by these workers. For the industrial contract workers alone (leaving those in commerce aside), the former came to around one-third of a million yuan under both the pre- and post-1972 payment systems, an amount almost equal to the total budgeted by the state for relief to poor communes in 1978. The latter is harder to estimate, but if each contract worker were only remitting 100 yuan per year—less than one-fourth of average wages—this would come to a total flow into the rural household sector of almost 1 million yuan. This amounts to about 40 percent of all state assistance to agriculture in 1978. Of course, against these contributions to the rural collective and household sectors must be weighed the income lost to the rural sector from the loss of these workers' labor; but given the high labor/land ratio and low level of marginal labor productivity in Shulu agriculture, this must be minimal.

A second benefit to the rural areas from contract labor has to do with the skills and contacts contract workers make in town and bring home to the countryside. These can often make the difference between economic success or failure for the rural enterprises that are so crucial to the overall rural economy of a locality, though evidence on their contribution is usually anecdotal. Third, the rural collective sector is spared the suffering and potentially disastrous effects of uncontrolled outflow of peasant labor to towns, a historical phenomenon which has made recent crises in the rural economy, such as that which occurred after the Great Leap Forward, worse.[17] Fourth are the redistributive effects of the contract labor system among rural units. These operated in various ways at different times. The pre-1972 payment system benefited poorer teams more because they were receiving the same amounts per worker from the urban employers as the richer teams, but were in turn paying the workers in less valuable workpoints, thereby retaining more for their collective funds. Even the post-1972 payment system made some allowance for intrarural redistribution by stipulating that poor teams would receive slightly more of their workers' wages than richer teams. Moreover, Shulu officials said that poorer teams are given preference in the allocation of contract work slots. Of course, against these redistributions must be balanced the effect of contract work in opening up new inequalities between the contract workers and their fellow peasants.[18]

It appears, then, that contract work has contradictory effects on urban-rural relations. It sets in motion flows of considerable sums from the urban areas into the countryside, flows which would not exist at all were townward migration either prohibited or uncontrolled. Under it, peasants have the chance to earn much higher incomes than they could otherwise. It has other salutary effects on the rural areas too. Yet, it provides for clear inequality between peasant contract workers and regular urban workers in pay and other conditions of employment. It can do this precisely because it rests upon the structural cleavage between peasants and workers in China: specifically, contract workers will do the same work as urban workers for lower wages and on more unfavorable terms of employment because of the severe obstacles to getting higher-paying urban work any other way.

In other words, contract labor rests firmly upon the structural cleavage between urban and rural people, even as it effects, at least in Shulu, some significant financial flows across that cleavage into rural areas. Moreover, to the extent that it permits state-regulated movement of labor power from countryside to the city in places where it is needed, contract labor prevents the state from having to dismantle the household registration system. In this sense, contract labor may actually be seen dialectically as reinforcing or at least preserving the urban-rural cleavage at its point of greatest vulnerability.

Yet, at the same time, contract labor may be producing new class forces that will bring pressure against the household registration and contract labor systems themselves. Contract workers form a distinct stratum of the peasantry. They remain tied to it by the rural collective units' continuing claim on their labor, their occupational futures, and a portion of their income, as well as by the fact that their families usually stay at home in the village. Yet their nonagricultural occupations, their opportunity in many cases to reside in a town (albeit on an irregular or subaltern basis), and their higher incomes set them clearly apart from the peasantry. These factors bring them within the closest possible proximity to the urban working class, but they are denied equal status with members of that class in terms of income, job security, fringe benefits, and access to opportunities for upward mobility for themselves (such as job promotions) and for their children (such as superior schooling and assignment to urban employment). The contract workers' objective situation engenders a clear interest in removing the restrictions embedded in the household registration and contract labor systems, so that they can become regular workers.[19]

There is evidence that this objective interest has produced a subjective one among the contract workers themselves. Contract workers were, after all, among the most radical participants in the Cultural Revolution, which they used as an opportunity to express all manner of grievances concerning their lower pay, less secure tenure, and greater vulnerability to attempts by managers to enforce various kinds of labor discipline on them.[20] In Shulu contract workers in the important Fur and Leather Tanning Factory attempted to utilize an official policy in effect up to 1972 to have their household registrations changed from rural to nonrural. This was so controversial that the entire policy that permitted such changes was reversed; probably as a compromise, the payment system for contract workers was changed in that year so that the workers could retain more of their salaries. In a sense, this amounted to a trade of greater urban-rural balance for retention of the basic urban-rural cleavage embodied in the household registration system. In short, the system of contract work, which is based upon and even reinforces the household registration system and through it the worker-peasant cleavage, has created a social stratum that strongly opposes continued maintenance of the cleavage and the regulatory systems associated with it, and which has already demonstrated its capacity to act in concert to express its opposition in political action.

### Conclusion: Balance and Cleavage Between Urban and Rural in Shulu

This chapter began by positing an analytical distinction between balance and cleavage within the urban-rural *problematique*. This distinction helps to sort out some of the apparently contradictory phenomena that have resulted from China's socialist development on a national level: for example, that urban-rural differences have narrowed while at the same time peasants are strictly denied access to the urban residence, regular industrial work with all its fringe benefits, and more varied and open channels of mobility that have been the birthright of most urban dwellers. China appears to have made progress in greater urban-rural balance, but at the same time it has done little if anything to overcome or bridge the structural cleavage between urban and rural at the level of social relations, political organization, or mode of production. The question posed here is whether this general picture holds up in the study of a single locality—in this case, a county—with its own urban-rural character.

In at least two of the areas examined here, the answer appears to be affirmative. First, urbanization in Shulu has been proceeding by reaching out into the rural areas rather than in isolation from them. The major urban center of Xinji has developed rapidly on a material base of expanding administrative, industrial, and commercial activity. But at the same time, smaller towns scattered about the Shulu countryside, building upon their historical functions as centers of administration and/or commerce, have also been growing, partly with the encouragement of county government, which has used them as subcenters of administration and governance. Urbanity is being brought more closely within the reach of rural people by proliferation of these small urban places. Political reforms now underway to strip the people's communes of their political functions and vest them once again in township governments may be expected to consolidate the development of these towns and provide them with a firmer political base than they have had since 1958. In this one respect, at least, state activity, so often seen as a centralizing force, has within Shulu assumed a rather different character. The distance between town and country is being reduced.

Yet, this tendency raises several important questions concerning the town-country *problematique*. In economic terms, the growth of small rural towns may hold out the prospect of bringing modern economic benefits (relating to employment, income, and consumption) within closer reach of the rural population. But will it also usurp certain economic developments, especially in rural sideline and industrial activity, that have grown up in the grassroots collective units in the past twenty-five years? Certainly some of the emphasis of recent economic reforms on efficiency, rationalization, standardization, and integration would promote the latter tendency, as would the political power now being vested in the new township governments. In Shulu, one prosperous commune-run agricultural machinery factory was already in 1979 anticipating its integration as a subunit within a county- or prefecture-run network of enterprises, with most certain implications for its future growth and financial and entrepreneurial autonomy. Likewise, in social and sociological terms such towns hold out the promise that rural residents can be provided with more and better medical care, education, cultural life, and opportunities for upward mobility. But will the elaboration of these more localized nodes of social and cultural services be used to justify the closure of small rural schools, clinics, hospitals, mobile projection teams, cultural troupes, and the like? Will opportunities for some peasants to make leaps in social mobility to employment and life in small towns expand at the expense of chances for smaller steps (perhaps for more people) within the rural collective sector—for example to nonagricultural jobs in collective enterprises or administrations? In political terms, the hiving off of political power from the communes to the township governments is being justified on the grounds that it will reduce political interference in the rural economy. Will it also reduce the political voice, influence, or power of the rural collective sector in (or against) the state? Will local political participation become impoverished, political conduits from the grassroots upward broken, and the hand of the state, which used to rest at the county level and only partially at the people's commune, actually be strengthened vis-à-vis the rural areas now that it will repose in unalloyed form at the lower level of the township? Returning to the distinction between balance and structure, as the size of the urban-rural gap diminishes, in this case in terms of distance, might not the social, cultural, economic, and political cleavages between them grow more chasmal?

Second, there is the contradictory phenomenon of temporary contract labor. On the one hand, it seems to offer some significant benefits to both the urban and rural areas it affects. The contract workers receive higher incomes than they had on the farm, escape the drudgery of farm labor, and gain industrial skills and experience. The rural collective units receive a share of the contract workers' incomes and are relieved of excess labor capacity while still retaining the option to recall that labor force if need be. They may also benefit from skills and contracts brought home by the contract workers. The urban employers get a supply of labor that costs less than regular workers and which can be employed and managed more flexibly. Urban governments are spared many of the costs of social reproduction of this labor force, such as housing, education, and urban services, and the towns can avoid overcrowding and the problems of squatter settlements. But there are drawbacks too, especially on the rural side. Contract workers are underpaid relative to their urban co-workers for often identical work, the conditions of their employment are less secure, and they do not receive the same benefits or entitlements such as housing, health care, educational opportunity, or claim on urban employment for their children. The peasantry in general suffers continued restriction of freedom of migration. Inequalities among peasants employed in contract labor and those remaining in agriculture grow. How all these balance out in terms of a final urban-rural accounting sheet is impossible to say, among other reasons because some of the major factors are not quantifiable and others are not measurable. We shall have to content ourselves with the more modest if somewhat unsatisfying conclusion that the contract labor system seems to strike some sort of rough if ultimately uneven balance (though in which direction the unevenness goes is not clear). The system does not indisputably benefit either rural or urban people at the others' expense.

Yet, contract labor can only strike such a balance as it does because it is based upon the continuing cleavage between urban and rural people, in the form of the household registration system, which divides the Chinese people into formal urban and rural categories. These in turn embody continuing inequalities in income, access to social and collective goods, cultural milieux, occupational constellations, and mobility opportunities. Moreover, contract labor seems actually to reinforce the household registration system, and through it the structural cleavage of urban and rural, at the point where it would otherwise be most subject to pressures to transcend it: namely, where peasant labor is needed for industrial growth.

Evidence from Shulu county, a rather ordinary place, suggests that in at least some areas of urban-rural relations—the pattern of urban or town settlement, the regulated movement of rural labor into urban employment, and certain very specific, limited areas of state budgetary activity—some measure of balance has been struck. Yet, little progress has been made in overcoming the cleavages between urban and rural areas in terms of the sturctures of economic, social, and political life. If anything, some of the balance has resulted precisely from the maintenance of these cleavages.

The persistence of structural cleavages between town and countryside during the transformation from precapitalist to capitalist modes of production and social formations (with their associated modernizations of the productive forces) has proven just as, if not more, explosive as the persistence of imbalances in urban-rural incomes and standards of living.[21] China cannot forestall forever facing this problem, which runs so deep in its history and culture and which forms today so important a component of its class structure and hence its socialist "laws of motion," whatever they may be. The leaders of China's revolution have been aware of this, though they have differed on how to deal with it. The Maoists tried to attack the urban-rural cleavage directly, while the post-Mao leadership seems to feel that the development of the productive forces under some basic but reduced scale of state regulation and corrective action will eventually take care of things. Meanwhile, China's peasants are being asked to content themselves with some improvements in the area of urban-rural balance, and to forestall attacking the urban-rural cleavage directly. This may work for the short term, but it has real risks, not only for the political position of the present leadership but also for the smoothness of China's progress toward becoming a modern socialist country.

## Notes

1. Suzanne Paine, "Some Reflections on the Presence of 'Rural' or 'Urban Bias' in China's Development Policies 1949-1976," *World Development* 6, 5:693-707.

2. William Parish and Martin Whyte, *Village and Family in Contemporary China* (Chicago: University of Chicago Press, 1978), pp. 52-54.

3. The estimates for China come from several sources: Parish and Whyte, *Village and Family,*; Benedict Stavis, "The Standard of Living in Rural China, 1978-1979," unpublished ms., November 1980, p. 60; William Parish, "Egalitarianism in Chinese Society," *Problems of Communism* (January-February 1981), p. 41; William Parish, personal communication. Michael Lipton sums up the data from other third world countries with an estimated urban-rural income ratio of over 3:1, running in some cases as high as 6:1. See Lipton, *Why Poor People Stay Poor: Urban Bias in World Development* (Cambridge: Harvard University Press, 1976), p. 150.

4. Xue Muqiao, *China's Socialist Economy* (Beijing: Foreign Languages Press, 1981), p. 147 ff.

5. For pre-1975 data, see Nicholas Lardy, *Economic Growth and Distribution in China* (Cambridge: Cambridge University Press, 1978), pp. 175-78.

6. Rhoads Murphey, *The Fading of the Maoist Vision: City and Country in China's Development* (New York: Methuen, 1980), p. 22. Reference is made to the large and medium-sized urban factories dubbed as "collective" in the Chinese system, such as those under ministries like the Second Light Industry Ministry, those known as "county-run collective industries," and those run by county governments directly, whose profits go into county extrabudgetary funds. These are discussed in detail in Vivienne Shue, "Beyond the Budget," *Modern China* 10, 2 (April 1984) and Marc Blecher et al., *Town and Country in a Developing Chinese County: Government, Society and Economy in Shulu Xian*, forthcoming. Shulu was not atypical in this regard, at least for its area. In 1978, per capita GVAO in Shijiazhuang prefecture was 220 yuan, while per capita distributed income was 84.31 yuan.

9. G. William Skinner, "Marketing and Social Structure in Rural China (Parts I-III)," *Journal of Asian Studies* 34, 1-3 (November 1964-May 1965), pp. 3-43, 195-228, and 363-99.

10. Ibid., p. 6.

11. Ibid., p. 7.

12. See Vivienne Shue, "Beyond the Budget," and Blecher et al., *Town and Country.*

13. Specifically, it includes on the revenue side the proceeds from the agricultural, income, and unified industrial and commercial taxes, and the profits of nine local state-run plants. On the expenditure side it includes monies allocated by the state for

agricultural assistance, support for rusticated youth (by now probably obsolete), emergency welfare aid, government administration, education, cultural programs, broadcasting, scientific work, health care, and sports.

14. The question of the representativeness of 1978 arises, especially in view of the large expenses involved in financing a massive water conservancy project built in southern Shulu in the winter of 1977–78. In 1978 expenditures for "support for agriculture" increased at about the same rate as they had been since 1976, and at slower rates than in 1973 and 1974. Much of the water conservancy work was financed by transferring funds from what probably would otherwise have gone to the rural areas anyway under the subcategory of "support for communes"; 1978, therefore, does not appear to have been particularly unusual in terms of its urban-rural balance *in official state budgetary expenditures*. Lacking any historical data on profits of the eighteen industries under the Second Light Industry Bureau, which we would need to estimate the urban (and by extension the rural) contribution to the income tax, we cannot say too much about the revenue side. It is our sense that these eighteen industries had been growing rapidly up to 1978, probably not at a rate very different from that for commune- and brigade-run enterprises. If so, then the relative contribution of the rural and urban areas to the income tax would not have changed significantly in the immediate past. We do know that agricultural tax revenues had remained pretty constant since the mid-1960s (except for a brief dip in 1977, no doubt a result of tax abatement in the wake of the 1977 flooding), and that industry and commerce tax revenues had been rising very rapidly since 1967. Taken together, these estimates and fragments of information do not seem to point to any major urban bias in the official state budget in the decade or so previous to our study. And in view of the enormous size of the industry and commerce tax, it would appear that if anything the 1978 balance lies somewhere at the end of a long-term process in which the official state budget has been increasingly tilted toward the rural areas.

15. Although there are some ambiguities in the materials, we take these figures to represent new loans extended each year rather than annual outstanding loan balances.

16. We have, unfortunately, had to resort to using 1979 data here; we were not given data for 1978. Also note that we have no data for any savings deposits in accounts of urban industrial or commercial enterprises.

17. For example, see Tang Tsou, Marc Blecher, and Mitch Meisner, "Organization, Growth and Equality in Xiyang County, Part II," *Modern China* 5, 2 (April 1979):140–54 and *passim*.

18. These complex effects are discussed in Marc Blecher, "Peasant Labor for Urban Industry: Temporary Contract Labor, Urban-Rural Balance and Class Relations in a Chinese County," *World Development* 11, 8 (August 1983).

19. It may be that the peasantry as a whole shares this opposition to household registration and maintenance of the peasant-worker cleavage. Their lower income and standard of living and their more toilsome conditions of work provide an *objective* material base for such opposition. Yet it is not at all clear that most peasants would have an objective interest in eliminating the contract labor system. In view of the high rates of urban unemployment in China, most peasants seeking urban work would have difficulty finding it; many of those who did might end up in a heteroclite informal labor market doing periodic, insecure, low-paying work. Of course many might still want to be able to take their chances. But for those staying behind (excluding family of those who found lucrative urban employment), eliminating the contract labor system would mean the loss of collective revenues accruing from the collectives' share of contract workers' wages, and also loss of a claim on the contract workers' labor in emergency situations. Certainly this would be opposed by those most committed to and dependent on the rural collectives, such as the poor and some rural collective unit cadre. The precise lines of cleavage on this issue of course have a subjective component which cannot be discerned from even the most precise analysis of the objective conditions.

The point here is simply that there is no necessary reason to suppose that the peasantry as a whole would favor the elimination of the contract labor system and its replacement by unrestricted movement of rural labor to urban employment. Much more likely, the question would be controversial within the peasantry, one possible line of cleavage being between the strata of contract workers and some of those who continue to be employed in agriculture.

20. Hong Yung Lee, *The Politics of the Chinese Cultural Revolution* (Berkeley: University of California Press, 1978), pp. 130–32.

21. Barrington Moore, *Social Origins of Dictatorship and Democracy* (Boston: Beacon Press, 1967).

# 10

# Taitou Revisited: State Policies and Social Change

# Norma Diamond

The call for modernization, diversification, a balanced economy, increased production, and attention to local conditions is nothing new in the various documents, speeches, and debates concerning Chinese agriculture. But such calls took on a particular urgency in 1980–81 as they admitted to the recognition that from 1958 to 1976 disappointingly little progress was made in most rural communities and that for some units production had not only stagnated but had in fact retrogressed. These failures, it is clear, cannot be primarily attributed to the supposed innate conservatism and backwardness of the peasants in the teams and brigades that make up China's rural villages. They also are failures that stem from bureaucratic misdirection, from lack of understanding, and from rigidity in policy-making at levels considerably higher than the local community and its peasant constituents.

This chapter is a microstudy of one particular community over time. Taitou Brigade is a village in Shandong province in East Central China. It was originally the subject of a 1945 monograph by Martin Yang, a sociologist and native of the village.[1] I was able to conduct four months of field work in this same community during 1979–80, collecting oral histories, interviewing households about present-day life and the recent past, and copying economic records. During that same year I visited several neighboring villages and spent some weeks in one of Shandong's "model" brigades, gaining information that helps put Taitou's record of economic progress in perspective.[2]

By some standards, Taitou has done rather well. Its 1979 collectively distributed income of 176 yuan was much higher than provincial or national averages in the same year. Its annual growth in income and crop yields since the mid-1960s has also been much greater than the national averages (see table 10.1). Nevertheless, compared to its

potential, this growth could be described as only modest at best. Within its area it was only average, and the great progress of neighboring model areas that I visited, sharing initially similar conditions, suggests some of the growth that might have been realized with alternative sets of government policies. It was not the failings of village leaders and the peasants at large that explained Taitou's modest growth but particular state policies and the interpretation of these policies by county and prefectural officials. The character of bureaucratic intervention did much to shape the nature of village development.

Table 10.1

## Income and Productivity

|  | 1962 | 1966 | 1970 | 1974 | 1978 | 1979 |
|---|---|---|---|---|---|---|
| Income (Y) Per capita | 39 | 70 | 62 | 129 | 132 | 176 |
| Work day value | .49 | .58 | .45 | .72 | .71 | .75 |
| Grain dist. kg. per capita | n.d. | 216 | 190 | 244 | 248 | 295 * |
| Grain yields (kg./ha.) | n.d. | 2,369 | 3,028 | 3,462 | 3,085 | 5,696 ** |

*The 1979 figures seem too high but may be a result of double cropping on corn and sweet potatoes, or may be preharvest estimates.

**Grain yields include sweet potatoes divided by a conversion factor of five. In the distribution, sweet potatoes account for between 40% and 55% of the per capita grain until 1979 when they fall to 20%.

## Physical Setting and Historical Background

Taitou lies a few miles from the seaside, twenty miles across Jiaozhou Bay from the city of Qingdao with its population of over a million. It lies in a flatlands area, ringed by mountains. It is a natural village, which in 1979 had a population of around 960 persons on 86.8 hectares of cultivated land. Since the late 1970s, it has been a part of Huangdao district (*qu*), which in turn is a part of Qingdao municipality (*shi*). Until the late 1970s, however, it was administratively and economically a part of Jiaonan county (*xian*), which in turn was the southernmost boundary of Changwei prefecture. For Taitou and its neighbors the nearest railway station is Jiaoxian, a trip of some sixty kilometers to the north over poor mountain roads. From there it is another four or five hours by train to Weifang, Changwei's administrative city with perhaps a quarter million population. Thus, while physically close to a major city and international port at Qingdao, for three decades Taitou was consigned to the administrative periphery of its prefecture. The core of that prefecture, centered on Weifang and surrounding counties, particularly Anqu, prospered, with the prefecture being named as one of two provincial success stories at the Third National Conference on Agricultural Mechanization in early 1978. But the periphery of the prefecture shared in few of these remarkable gains.

Taitou's separation from Qingdao municipality began in 1945 when it was liberated by the People's Liberation Army while, with help from the American navy, Qingdao municipality remained in the hands of the Nationalists until 1949. In the intervening years, prior to the establishment of the People's Republic of China, land reform was completed in the region and informal mutual-aid teams encouraged. For these and perhaps other reasons, the Jiaonan region was assigned to Changwei prefecture, which similarly had been liberated early in the Civil War, rather than to Qingdao municipality.

Table 10.2

**Pre-Land Reform Distribution of Households and Land**

|  | No. of households | % of population | % of land | Amount of land per capita (hectares) |
|---|---|---|---|---|
| Landlord | 1 | 1% | 4% | .48 |
| Rich peasant | 5 | 5% | 18% | .47 |
| Middle peasant | 90 | 59% | 57% | .14 |
| Poor peasant and hired labor | 44 | 34% | 21% | .09 |
| Total | 140 | 100 | 100 | .14 |

In 1945, Taitou was a community of small landowners, with middle peasants, artisans, and petty merchants predominating. Members of some households were deployed into urban (working class) jobs. There was only one small landlord family resident in the village, no absentee landlords, and relatively few tenants or poor peasants compared to other communities (see table 10.2). With so little land and equipment to distribute from the landlord household, rich peasants and even a few middle peasants had to give up land, tools, and draft animals to equalize holdings. The level of education was relatively high: in 1945 it could claim 97 literate persons, most of whom came from middle and poor peasant households. Within its standard marketing area of some sixty-eight villages, it had one of the five schools.

Following land reform, cropping patterns remained similar to what Yang described for the earlier period. Spring wheat, in this area a low-yield crop, and barley were sown for the early summer harvest, while millet, sweet potatoes, and to a lesser extent corn and sorghum (kaoliang) were the autumn harvest crops. Peanuts and soybeans remained important both for household consumption and commercial sales. Some of the land was used for cotton, and some to grow a wide variety of vegetables both for home use and sale. After the liberation of Qingdao in 1949, private marketing of vegetables to that city resumed for a time.

In later calls by the state for village reorganization, Taitou's response was about average. It responded positively to the 1952 call for mutual-aid teams, with 109 of the 154 village households participating. The response to the 1954–55 call for lower-level agricultural producers' cooperatives was slower. Only eighteen of the households, representing a consolidation of two mutual-aid teams plus a few other households, agreed to join voluntarily. The following year, the matter was no longer voluntary: by government decree the whole village became a single higher-level agricultural producers' cooperative. The continuation of small-group voluntary cooperation was impossible, for only the new cooperative could make grain sales to the state or obtain fertilizer, new seeds, and credit. Within the cooperative the work force was organized by household into one of six teams. Households could indicate which team they wanted to join, but the final decisions were in the hands of the team leaders (all party members) or the leadership of the cooperative. A small amount of land, about 7 percent, was kept out of the team-assigned fields and allocated to households as private plots.

Barely two years later, in 1958, Taitou found that it was no longer a higher-level APC but a brigade, and a part of Xingan Commune along with some sixty-seven other villages. As one informant said, it didn't really seem to make that much difference at the outset; it was more a change of name, "like the Eighth Route Army changing its name to People's Liberation Army." The unit was already divided into teams (*shengchan dui*), and in Taitou these remained the basic production and accounting units. Teams were asked to put similar amounts of land into each crop, with production remaining varied during the initial years of the commune system. That is, the teams continued to grow cotton, soybeans, peanuts, and specialty crops like melons and peppers in addition to the staple grain crops of wheat, millet, sweet potatoes, barley, corn, and sorghum.

Taitou appears to have moved through these very radical changes in land-holding patterns and labor organization rather smoothly compared to many places in China. And through these years, the bureaucratically organized farming system continued to add to the prosperity of the village. During the 1960s, the new commune organization was used to mobilize labor from several villages along the Taitou River to dig a new river channel, wider than the original one, and divert the river into a new course. Though this put the river at a greater distance from Taitou village, it reduced the dangers of flooding of fields and homes and created a new strip of fertile land for fruit orchards. The commune also lent assistance in the digging of new wells and construction of a water-storage tank, and it distributed and instructed on the use of new chemical fertilizers and insecticides. In the early 1960s the commune and the bureaucracy also helped introduce a new strain of corn to replace millet, sorghum, and barley as household grains. This corn was well received, and it became an important part of the household diet until the early 1970s. These were all helpful aspects of state intervention. Despite the haste with which collectivization was accomplished, Taitou's people seemed strongly supportive of a socialist path to development. Despite setbacks, which will be discussed below, the past three decades have brought a higher standard of living and clear benefits such as access to medical care, expansion of education, some mechanization of agriculture, and the framework for the organization of human effort to level and consolidate fields and move rivers. Even those with "rich peasant" labels regard themselves as now markedly better-off than they were in the 1940s, eating better and being able to purchase consumer items that were rare or unheard of in earlier times.

## Grain-First Policies

In the late 1960s, however, state policy began to turn in a less helpful direction, emphasizing the production of particular grains and higher production targets that incurred higher production costs. Income stagnated and there was a decline in the quality of local diets. The situation Nicholas Lardy has reported for the nation as a whole elsewhere in this volume was repeated in Taitou. The directives that came down to this backwater of the Changwei administrative area did not take local conditions into account. The area has a maritime climate, warmed by the sea and sheltered by the surrounding mountains. It thus has milder winters than Shandong's inland areas. In winter, rain is more likely than snow, and when the temperatures go below freezing there is likely to be no snow cover to protect the wheat seedlings. The summers are milder than in the inland areas (there is about a 10-degree difference between Jinan and Qingdao during the summer), but there is often heavy fogging in the warmer months. Burn-off of the fog may not occur until three or four hours after sunrise. Because of the coastal location, much of the soil is of a sandy quality, well-suited to sweet potatoes and peanuts, workable for millet which is a hardy crop, and less suitable for wheat.

Despite this, Taitou was instructed to emphasize wheat and corn while phasing out millet, sorghum, and other grain crops regarded as "low yield." In 1970, a new strain of corn was made mandatory—a

Table 10.3

**Land Use and Productivity—Grains**

|  | 1966 | 1970 | 1974 | 1978 | 1979 |
|---|---|---|---|---|---|
| 1. Area (hectares) | | | | | |
|   a) Wheat | 33 | 43 | 52 | 60 | 52 |
|   b) Corn | 23 | 10 | 20 | 42 | 36 |
|   c) Sweet potato[*] | 33 | 38 | 26 | 18 | 25 |
|   d) Millet | 2 | 6 | 3 | 1 | — |
|   e) Peas | 12 | 5 | 7 | 1 | — |
|   f) Totals | 102 | 102 | 108 | 122 | 113 |
| 2. Yields (metric tons per hectare) | | | | | |
|   a) Wheat | 1.3 | 1.8 | 2.6 | 2.2 | 4.8 |
|   b) Corn | 2.7 | 5.3 | 4.7 | 3.7 | 7.4 |
|   c) Sweet potato[*] | 3.8 | 4.3 | 5.0 | 4.8 | 5.0 |
|   d) Millet | 2.7 | 1.0 | .7 | 1.9 | — |
|   e) Peas | .6 | 1.6 | 1.9 | ** | — |
| Average per ha.[†] | | 2.4 | 3.0 | 3.7 | 3.1 5.7[#] |

[*]Sweet potato yields already divided by standard conversion factor of five.

**No data available.

[†]With rounding, totals and averages may differ from preceding figures.

[#]The 1979 figures seem too high but may be a result of double cropping on corn and sweet potatoes, or may be preharvest estimates.

variety which was still in use in 1980. Though the new hybrid produced spectacular results in the first two years of its use, these yields proved difficult to maintain (see table 10.3). Continued high yields required increased inputs of expensive fertilizer. More seriously, from the peasant's point of view, the taste and texture of the corn was better suited for animal fodder than human consumption. That may even have been the intent of the developers of the hybrid, but from 1966 to 1976, private (household) raising of pigs was discouraged and carried out on a very small scale, so that the potential fodder-value of the corn went to waste.

In 1966, Taitou's grain quota was 31,500 kilos; in 1970, it had risen to 67,574 kilos. To meet the demand for increased grain deliveries to the state, Taitou was instructed to put still more of its land into wheat and corn and pay more attention to double cropping. Land could not be left fallow, as had been the practice in the past. Area allowances for soybeans and peanuts had been reduced earlier; now they were further reduced for "coarse grains" (including sweet potatoes) and completely abolished for highly profitable cotton (see table 10.3).

These demands were particularly troublesome in that local wheat yields continued to be low in comparison to the more northern sector of Changwei prefecture. In the model brigade of Shijiazhuang, Anqu county, which has good irrigation, the 1978 wheat yields averaged 7.5 tons per hectare. The norms for the entire prefecture were undoubtedly lower, but even so it is safe to venture that Taitou's natural conditions for growing wheat put it in a position where at best it could achieve only half the results of more favorably situated places. The administrative response was not that they reduce the emphasis on wheat and shift to other crops. On the contrary, the response was that they should try harder, use more fertilizer, more man hours, and more land.

With the insistance on high state grain deliveries and wheat production, the grain diet improved only marginally prior to 1979 (see table 10.4). Peasants prefer to eat wheat, and they did eat more in the 1970s than a decade earlier. But their major dependence continued to be on the sweet potato. When state purchasing requirements were relaxed starting in 1979, they rapidly switched more of their consumption into wheat, allocating the collectively distributed corn to fodder for the expanding pig-raising industry, while sweet potato continued to be used for both fodder and human consumption. Also prior to 1979, the production of miscellaneous grains not desired by urban consumers was severely cut, and local consumption decreased. Millet, seen by the peasants as a good food for children, pregnant women, and the elderly, and simply the accustomed diet of the older generation, was less than ample. Some people turned to growing millet on their private plots in place of vegetables, and the free-market price for millet rose above that of wheat or corn. Soy flour and barley, which was formerly mixed with soy flour to make a daily porridge, disappeared from the diet.

Table 10.4

**Average per capita Grain Distributions by Year (kilos)**

|                    | 1966 | 1970 | 1974 | 1978 | 1979 |
|--------------------|------|------|------|------|------|
| Wheat              | 37   | 59   | 65   | 60   | 110  |
| Corn               | 59   | 49   | 80   | 78   | 125  |
| Sweet potatoes*    | 120  | 82   | 99   | 110  | 60   |
| Total              | 216  | 190  | 244  | 248  | 295  |

*Converted at 5-10-1 ratio.

There was a similar decline in other items that added to the variety and nutritional value of the diet. Soybean acreage and distribution was cut in half, reducing the yearly distribution to 4 or 5 kilograms per person. Increased private poultry and egg production in the late 1970s compensated for the decline in this source of protein to some extent. But without the beancurd, which contains amino acids that facilitate extracting the maximum nutritional value from a grain-based diet, people had to eat more grain than before. Peanuts, important for proteins and cooking oil (as well as being a cash crop in the past), remained scarce. The amount distributed was about 7 kilograms per year, which was insufficient to supply the estimated need of 5 kilograms of cooking oil per person.

Vegetable production was also ignored, except on the private plots. The only vegetable allowed on collectively owned land was turnips, grown on less than 3 hectares of land each year, and providing a distribution of between 25 and 50 kilograms per person. White potatoes were defined as vegetables (*cai*) and grown on private plots along with cucumbers, eggplants, green peppers, string beans, tomatoes, squash, cabbage, and a few minor vegetables. Salted or sun dried, many of these vegetables were available throughout the year. The village does not grow much cabbage—the soil is not really suitable, and it is this feature that has apparently led to the bureaucratic judgment that the area is unsuited for vegetable production and therefore not to be assigned to provide vegetables for the urban areas. I tried to discuss this with one of the Huangdao district leaders, praising the various local produce, but somehow the conversation kept coming back to the superiority of the cabbages grown around Jiaoxian. At that time, Jiaoxian supplied most of the vegetables to the growing town of Huangdao, despite transportation costs.

Units farther north like the model Shijiazhuang Brigade, which scored successes in the growing of grain (thanks to climate, soil, and availability of irrigation), were "rewarded" throughout the late 1960s and 1970s by being allowed to move some of their land into other, more valuable crops. Many of the villages in Anqu county, for example, grew tobacco, with yields and sale prices five to six times those of grain. They were encouraged to grow melons, another high-value crop, for state sale. Because of the high labor demand that accompanies tobacco processing, some of these units gave up private plots and were permitted to engage in diversified vegetable farming on collective land, both for distribution purposes and for outside sale. In contrast, units like Taitou, which get low grain yields, were not allowed to make up the income difference by producing high-value crops to compensate for low grain productivity. To add to the paradox, advanced units were rewarded with access to farm machinery, as a gift or as rights-to-purchase, further boosting their production levels and opening the way both to higher income and to further acquisition of machinery or rights to engage in high-value crop production and side industries.

Taitou's agriculture was only barely mechanized, despite its claim to own "twenty pieces of machinery." These included two 25-horsepower tractors and three 12-horsepower tractors, used mainly for transport work. They owned no attachments to make these machines usable for seeding or harvesting, though some of the ploughing was mechanized. The three motorized threshers were shared among the five teams that now make up the brigade, but much of the work on the threshing grounds was done with traditional flails, rakes, and baskets. There was one machine for chopping fodder and sweet potatoes. The grain mill had two small milling machines. The oil shop, no longer in use in 1980, had a press, a crusher, and four smaller machines. The machine shop, an outgrowth of the blacksmithy of an earlier period, had a lathe, planer, and drill. The official count of twenty machines overlooked a brigade-owned sewing machine and a motor-run insecticide sprayer. Even with those oversights, it is clear that Taitou still relied mainly on animal and human labor power. If my account sounds a bit biased it may be because I arrived in Taitou fresh from the June harvest in the model Shijiazhuang Brigade. There I had watched mechanical reapers, threshers, and a combine harvester at work. In Taitou, the grain was harvested by hundreds of people equipped with knives, and it was threshed and winnowed with tools that go far back into Chinese history.

## Collective Sideline Policy

Taitou's economic development fell below its potential not only because it was forced to grow unprofitable grains but also because its links to the urban economy and the growth of industrial sidelines were stifled. This is particularly disappointing given Taitou's early ties to Qingdao city and the growth of the nearby port of Huangdao in recent years.

Because there was so little land per person prior to 1945, many if not most families had some members working in nonagricultural endeavors, often outside the village. This was true at all class levels. Rich peasants and the one landlord owned oil-pressing shops, grain mills, and a smithy in the village as well as shops in the nearby market town and even a biscuit factory in Qingdao city itself. Middle and poor peasants worked both locally and in Qingdao as carpenters, stonemasons, brick and tile workers, basket makers, peddlers, transport workers, and factory workers.

Some of these patterns continued after 1945. Craftsmen and other outside workers received proportionally less land at land reform and continued their nonagricultural jobs. The two oil-press shops and the blacksmith shop, which belonged to rich peasants, were reorganized as cooperatives and continued to operate. A small beancurd shop was also organized. In the mid-1950s, under the higher-level agricultural producers' cooperative, these shops were joined by a carpentry collective making farm tools, wheelbarrows, tables, desks, and stools as well as by a brick-and-tiles kiln.

In 1958, with the start of the commune, the village established a grain mill, a sewing group, and three communal kitchens, which are remembered to this day as providing good food at a financial loss. Freed from some of their household chores, women were mobilized for agricultural work and to search the nearby hillsides for potential sources of iron. For a brief time, the village worked its own iron mine, but the ore proved to be of low quality and the project was discontinued.

This burgeoning diversification was short-lived. The communal kitchens closed after the first year, and the sewing group lasted only another few years before being phased out. Moreover, the brick-and-tiles kiln reportedly found it increasingly difficult to market its products and was closed. The carpentry workshop members were told that they could not continue as a full-time enterprise and were reassigned to the agricultural teams. About half of the original group managed to be assigned to the same team, where they tried to continue their crafts specialization as a team-level sideline. They worked in this fashion until 1966, when they were attacked as "capitalist roaders." During the same period the beancurd shop closed as a result of declining soybean plantings.

With little encouragement and sometimes active discouragement of alternative sidelines, there was little growth in these other sources of income through the mid-1970s. The only exception was the sudden opening of possibilities for work connected with the building of the new oil port at Huangdao, starting in 1973–74. These opportunities, however, tend to illustrate the limitations rather than the possibilities for a place such as Taitou.

Starting in the mid-1950s, the migration of peasants into cities began to be severely restricted through new state regulations. Today, with continued unemployment problems in the cities, the policy of restricting peasants to the villages is maintained. Those Taitou residents who are working to build Huangdao do so not as individual laborers but as temporary contract workers paid through their home production team. They get none of the benefits accorded to urban workers and have no rights to permanent urban jobs. This sharp disjunction between urban workers and rural peasants is in sharp contrast to Taitou's pre-1949 situation.

For the time being, this temporary work adds tremendously to Taitou's income, accounting for most of the spurt of recent years. In 1970, Taitou's sideline gross income was 9,508 yuan, but by 1974 it was 91,246 yuan, even without the opening of new collective industrial sidelines, and by 1979 it was 135,966 yuan. This new income from contract work accounts for the doubling of workpoint values and collective distribution value. Taitou's peasants are construction laborers on new roads, residences, factories, and office buildings, laying bricks and stone, carrying these same materials, and doing other all-purpose jobs. Teams have assigned boats and carts, along with workers, to help move building materials from quarries to the building sites. But these work opportunities will pass in time. Huangdao port's population was already halfway to its intended 45,000 in 1979. Soon the town will have its own construction crews. And it has already been stated that the factories located there will recruit their skilled workers from other urban factories (including the relocation of Qingdao factories) and other workers, skilled or unskilled, from the urban unemployed.

Even though under the administrative control of the Huangdao urban district, villagers continue to hold a "rural residence" label that inhibits their chances of being assigned to an urban job. This is true for the young as well as for the old. A few young people have been allowed to attend the senior middle school in Huangdao, but the great majority of those who attend senior middle school do so in commune schools that do not prepare them for industrial jobs or for the competitive examinations now being used to select college students. The curriculum in brigade and commune schools is geared toward a life in agriculture. One district leader said to me, "Ninety-five percent of these children will continue to be peasants," as he explained why there was no need to change the curriculum. Despite the calls for modernization of agriculture, as one of the components in the four-modernizations program urged by Deng Xiaoping and his supporters, the Huangdao district leadership in 1979–80 did not foresee mechanization of agriculture in the near future, the growth of local industry, or any problem of a growing surplus labor force in the rural villages. The encouragement for creation of collective sidelines in the villages was minimal.

The chances for servicing the port with factory offshoots based in the villages may grow in the future but at present these prospects are only dimly perceived. Without them, Taitou's collective income will again decline by as much as 40 to 45 percent. The attempt to increase grain production in the 1970s drove production costs up from 25 percent of crop value in the late 1960s to 43 percent in 1978. It has since declined to 28 percent, due to increased grain-sale prices and greater local supply of fertilizer from renewed household pig production.

Collective pig raising operates at a loss in Taitou, and the fruit orchard accounted for only a little over 2 percent of gross income in recent years. Only a few additional opportunities have appeared so far. In the early months of 1980, Taitou was offered the opportunity to set up a printing shop that would print, cut, and assemble cigarette packages, and it willingly accepted despite inexperience in this line of work. Plans were then revised, again from above, to make the print shop responsible for printing the various report forms and circulars needed by rural units. Two retired masterprinters were hired by the district and assigned to Taitou, and a press and paper-cutting machine were purchased by the brigade at a cost of close to 20,000 yuan. By summer it was in operation, with some eight or nine young people being trained to set type and run the machines.

Elsewhere within Taitou's commune (Xingan) one of the brigades had received a contract to make coal shovels for a Qingdao factory, another was making valves, and in ten brigades women were doing embroidery, on a piece-work basis, for a Qingdao handicrafts factory. These are all post-1977 developments. Neighboring Xuejiadao Commune has been allowed to establish a rubber-tire factory and several of its fishing brigades have contracted with the Foreign Trade Bureau to raise mink. One of the fishing brigades began receiving assistance after 1977 from the Oceanography Institute in Qingdao, which helped it set up a laboratory for tank raising of abalone and beche-de-mer. In 1979, a few brigades were finally assigned to grow vegetables, on a small scale, for delivery to Huangdao.

These are not impressive developments when compared to the model Shijiazhuang. In 1979, brigade-owned industrial sidelines accounted for 52 percent of a gross income of 909,178 yuan. The parallel statement for Taitou is that industrial sidelines, which were primarily contract labor, account for 38 percent of a gross income of Y357,923 yuan. Shijiazhuang does have almost 200 more people than Taitou, and it spent a lot of money in 1979 buying additional agricultural and industrial machinery, but even so the year-end distribution averaged out to a value of 250 yuan per person, compared to Taitou's reported 176 yuan. Shijiazhuang's industrial sidelines are of three kinds: some, like the tailor shop and carpentry shop, essentially serve the brigade. Others, like the large-scale grain mill and brick-and-tiles kiln, serve surrounding communities as well. Most important economically are the brigade-owned enterprises that sell to purchasing agencies in Anqu and Weifang, and which produce plastic sacks, imitation leather or canvas handbags and totes, felt sleeping mats, plastic thermos bottles, and felt slippers. In late 1979 they had arranged to open a new factory producing parts for electric pumps. The inflow of money from this combination of agriculture and industry not only provided a high year-end distribution and allowed for the maintenance and expansion of production, it also subsidized a senior middle-school and a building program for new village housing that involved the labor of almost a third of the available work force. When asked what they could do with the surplus labor once the building program was completed, brigade leaders grinned and said they intended to introduce the eight-hour work day, and if possible reduce it to seven. The standard in rural units is a ten-hour day.

But perhaps it is unfair to compare Taitou or other brigades in Xingan with a provincial model. Shijiazhuang's neighbor, Tayulin, is regarded as only "average" for Anqu county: it has a modern poultry farm, a brick kiln, a paper-dying plant, and a workshop producing canvas bags. Reported 1979 collective distribution was 141 yuan in 1979, kept low in order to purchase agricultural and sideline machinery and expand brigade enterprises. Though its distributed income is lower than that of Taitou, its potential for growth is much better.

## Private Sidelines

State policy shaped life in Taitou not only through its control over grain and collective sidelines but also through its changing attitudes toward private sidelines. Private sidelines were preserved in the original collectives of the 1950s. Seven percent of the land was set aside for private plots, and households could raise pigs and chickens. The peasant market in Xingan continued to operate on its traditional five-day schedule.

This policy was sharply altered in 1966 when the market was closed and household-based enterprises actively discouraged. After 1976, the market was resumed and peasants once again allowed to develop private sidelines, which found a natural outlet through the market. In some of Xingan's villages, though not in Taitou, the peasants were encouraged to revive traditional home handicrafts: on market days it was possible once more to buy straw hats, straw sleeping mats in two or three different patterns, baskets, brooms and brushes, storage jars, and plain-fired black pottery. Individuals able to do repair work of various sorts were once again offering such services at the market. There was a special section where one could have shoes repaired, a watch or clock put back into working order, a pot handle replaced, or one's hair cut. Most of these activities involved male labor.

In Taitou through 1979, the relaxed policy meant only the reemergence of household raising of pigs and poultry and the introduction of rabbits. These ventures accounted for a large part of household income and were almost entirely carried out by women. In Taitou, the male labor force was involved fulltime in agriculture and in the transport and construction sidelines. In 1978, males worked an average of 361 days (ten workpoints equals one day), including extra credits for night shifts during the harvest seasons and field construction, trans-

port, and construction work during the slack seasons. Women's participation in collective agriculture and enterprises was considerably less:
the number of credited ten-point work days for women averaged 110.[3]
Their work was concentrated during the harvest seasons, in the fields
and on the threshing grounds. The further work of drying the grain
went unremunerated: it became an additional household chore for
women, older people, and children when the grain is distributed to the
households. Given the small land area relative to population, increased
participation by women in agricultural work would at best bring marginal returns or be redundant—unless more males were absorbed into
sideline work. In a similar vein, most of the sideline work excluded
women. They could work in the orchards, the new printing shop, the
grain mill, and oil-press shop but they did not drive tractors and carts,
sail transport boats, do heavy labor at the stone quarries, or serve as
masons and carpenters. Relatively few were assigned work as peasant
contract workers. It is worth noting that most of Shijiazhuang's
collective sidelines utilized female labor.

With the reopening of the market and approval given to household
sidelines, Taitou's women found ways of increasing household income.
Some of their work was carried out on the family's private plots, some
of it involved foraging the hillsides for animal fodder and herbs, or
going to the nearby seaside to collect shellfish. Some of the work went
on within the family courtyard. By 1979, most households were raising
two or three pigs a year for sale, feeding them on corn and sweet
potatoes from the distribution, additional foods from private plots and
the hillsides, and fodder distributed by the brigade based on sale weight
of the pigs. Most households reported a gross income from pig sales of
between 250 and 500 yuan. Additionally, the average household expected another 25 yuan yearly from sale of eggs and poultry, 20 from its
rabbits, and between 20 and 30 from shellfish. Despite the
bureaucracy's assertions that the area was unsuited for cabbages, several women reported earning 25 yuan over the year from the sale of
cabbages raised on their private plots. The other vegetables that were
raised were almost entirely for household consumption. Income from
household handicrafts was negligible. In the early part of the century
women spun cotton, but this sideline had long since died out. A small
sum could be earned around the time of Spring Festival by selling
papercuts, a craft in which a number of women are quite skilled, but no
other traditional handicrafts reemerged. The total for all these activities was somewhere between 350 and 400 yuan gross, on the average,
accounting for at least a third of household gross income.

In actuality, household pig production added a further source of income which was included in the collective distribution. The more pigs a household raised, the greater the amount of fertilizer it could sell to the collective. Households were credited with "fertilizer days" as well as "work days" in figuring the value of collective distribution due to each household. In 1979, some households were deriving as much as 40 to 50 percent of their collective income from fertilizer sales, though for most these sales represented only 25 to 35 percent of collective income.

It was apparent to virtually everyone I talked to that they stood to gain more in household total income by having one woman in the family devote her major energies to pig raising, participating in collective agricultural labor only at the busiest seasons. Her work would bring in credits for 200 to 300 additional work days for the household, far more than a woman's actual participation in collective agriculture could garner. Between 1978 and 1979, the number of fertilizer days increased from an average of 11,000 in each team to 15,500, suggesting a reasonable increase in private pig production as more families saw the advantages. Since higher levels did not authorize or encourage any collective sidelines that utilized female labor, and there was no need for their full-time engagement in agriculture, the liberalization of policies regarding private sidelines met with considerable support. Every household felt the need for additional income beyond what was available from the collective distribution, though in households that had only one full-time male worker women had to spend more time in the team in order to assure a sufficient grain distribution to the household.

A major expense that households had to deal with was housing. Since the mid-1970s, Shandong province urged a more consolidated, unified housing plan for villages. The houses should all be in one place, on ordered streets, and built to similar size and design. The details were left to the brigade to decide, and the brigade also decided how much it was able to subsidize the project. In Taitou, the specified design was similar to traditional middle-peasant housing—a one-story building with a central kitchen, and one or two sleeping rooms off to each side of the kitchen. It was basically a rectangle. The use of tile roofing, stone foundations and end-walls, and brick sidewalls were an improvement over much of the earlier housing. Depending on how much material could be salvaged from existing houses, the costs for a family ranged anywhere from 400 yuan to over 1,000 yuan for the house alone. Additional materials were needed for the courtyard wall and for storage buildings, privy, and pigsties that go into the courtyard. Some, but not all, of the labor costs were borne by the brigade.

One of the attractions of the new housing is that it is better construct-ed. Another, no doubt, is that when a number of other families have new houses you want one too. Yet another attraction is that a new house means more courtyard space for carrying out private sidelines. There have been an increasing number of household divisions in recent years, not just of brothers establishing separate households after marriage, but generational divisions in which the elderly couple stays in the old house while the married sons move into new ones. This increases the possi-bilities for raising additional pigs, poultry, vegetables, home tobacco, or whatever. With space at a premium, every little bit counts, particu-larly since population growth reduced the actual size of private plot allocations.

With the reopening of the market, and even the nature of Taitou's private sidelines and overall collective economy, women were drawn away from the village several days each month. The sale of eggs, poultry, and rabbits at the market is presided over exclusively by women. In the fresh-produce and shellfish sections of the market, both men and women are the sellers. The less frequent sale of pigs is left to the men; even though women are the primary caretakers, men are the ones able to get them to market and deal with the state purchasing agents or haggle with other peasant males. Similarly, the sale of fuel, tobacco, grain, most handicrafts, and repair services are presided over by males, but Taitou's free marketing is less involved with these items. Purchase of goods from the market (except for pigs) is mainly left to women, or to elderly males who are no longer fully active in the labor force.

The state policies that shaped Taitou's economic enterprises possibly did more to raise the status of women than the publicity campaigns surrounding the New Marriage Law of 1950 (which was not really implemented in Taitou until 1952–53) or the frequent exhortations to women to enter collective production and hold up their half of the sky. Not that the policies always had that intent, but they sometimes had that effect. With women doing most of the work connected with private sidelines and handling both the marketing of home-produced goods and the purchase of needed items, they came to have clearly recognizable importance within the household economy. In various interviews with peasant families and analysis of household budgets it was evident that women's work in the private sector equalled or surpassed the value of men's work in the collective sector.

In comparison to the traditional past, when women were regarded as less useful economically, women came to feel that they were entitled to their share of the better foods (including meat), to new clothing and pocket money, and to share in the decisions about the family's major purchases. When a peasant woman tells you that the boys get 10 yuan each as pocket money while she and her husband each get 25 yuan, or tells you how much her husband is allocated to spend on wine and cigarettes for guests, she is indirectly saying that household finances are no longer under male control. They are under joint control, and in some cases the wife seems to have a greater part of that power. Perhaps that is because the money she earns comes in throughout the year, not just at distribution or as a cash-in-advance loan from the collective, and because she has full knowledge of what the household's total financial resources are. Women informants on the whole seemed more knowledgeable than the men about the household's resources, its cash outflow, its savings, and its potential income.

State policies have brought about another change in family roles that bears mention. Because of the insistence on limiting migration from the rural areas into the towns and cities, some families are separated. Households in which the husband has obtained not only a city job but also city residence function as split households in which the children's residence status derives from the mother. There are a few cases in which women are married to men who work and live in Jiaoxian, Qaoxian, Qingdao, or Huangdao. The women and their children remain in the village, visited by husband/father on occasional weekends and holidays. There is no way for the man to bring his chiildren into town to attend the better city schools.

We should note here that at the time of land reform, household members were assigned a class status (poor, middle, or rich peasants) which was derived from that of the male household head. Nowadays, and for at least the past two decades, women born in peasant households retain "peasant" status even if they marry men with urban residence and urban job classifications, and the children derive their class status from their mother. While I was in Taitou, one young woman broke her engagement when her fiancee was allowed to enter a college. She argued that he would eventually be assigned to a town job and given urban-residence status. She did not want to spend the rest of her life struggling to raise the children by herself and living in her in-law's village. Another young woman, who married a worker in Huangdao, insisted on a matrilocal marriage, knowing that she and their children in future would not be able to live with him, and feeling that she would be best off remaining with her parents. Here, state policies intervene to create a permanent rural-urban division at the family level.

## Administrative Status and Rural Prosperity

I mentioned earlier that a small number of brigades in Taitou's commune and elsewhere in the Huangdao area were able to develop a more diversified economy, including nonagricultural sidelines. A few were regarded as advanced or "models." The question of "why some and not others" is dismissed by the present district leadership as a matter of village-level creativity, hard work, good political line, and the like. It is dangerous to generalize from such a small sample but I have the impression that the brigades given the most assistance and encouragement from higher up were those matching a certain blueprint of how development should proceed.

Such villages were very poor, or at least poorer than their neighbors. The model agricultural brigade in the Huangdao area is Guanting Brigade in Xuejiadao Commune, a village which at the time of land reform was notorious in the area as a poor tenant village. Of the hundred or so households, half were outright tenants, and another quarter lived by hired labor. Today they are the owners of the two stone quarries that supply Huangdao. They are permitted to raise kelp since they are near the sea, though like Taitou they are far enough from it not to be a fishing brigade. In 1970 they became the recipients of the first electric pumping stations in the area (Taitou as yet has none), and their grain yields, not surprisingly, are much higher than most places in Huangdao. In 1979, they had already begun operating a mink farm and had received permission to open a modern chicken farm. Half of the work force was involved in brigade-owned sidelines; very few were contract workers outside the community. Their origins are similar to the provincial model Shijiazhuang, which was also a tenant and hired-laborer village, most of its land owned by big absentee landlords.

Taitou, in contrast, began as a village of small holders, overwhelmingly middle peasant in classification. There were relatively few tenants, even fewer who were totally landless, no absentee landlords, and no big landlords in the village itself. It strayed further from the model of total poverty by being one of the few communities in the area to have a village school, in this case one that had begun under church auspices. Its higher level of education was of assistance in the entrepreneurial activities of the rich peasants and enabled some persons to leave the village permanently for skilled or professional jobs (including two university professors). Education also was important to the village artisans and others. Craft work, factory work in Qingdao, and small-trade ventures provided an alternative to petty farming or agricultural labor. In addition, the former presence of a Christian church in Taitou undoubtedly was a black mark against it, even if its active membership was relatively small. And during the Japanese occupation, the puppet government was headquartered in the market town of Xingan and recruited among the nearby villages for soldiers and clerks. Taitou's village history, written in 1974, records that at Liberation twenty of its people were targeted as collaborators with the Japanese. Before Liberation, the "model" village should not only have been poor, exploited, and illiterate but should have had a revolutionary heritage of resistance to the Japanese and the Guomindang. Only then did it seem to receive the kind of attention from higher levels that enabled it to make rapid progress and outpace those whose conditions at the time of Liberation were less dire.

There is the further question of Taitou's less-than-ardent enthusiasm for the formation of cooperatives, though its response was by no means unique in the general area. In contrast, Shijiazhuang with some outside guidance had brought 80 percent of its households into the lower-level cooperatives by 1955 and was cited as a county and provincial model.

Whatever the reasons, model villages generally had a long history in that status and continued to benefit from outside assistance and positive responses to requests for further development long after they ceased to be poor and disadvantaged in comparison to their neighbors. But there is another matter to be considered, and that is the degree to which the higher levels of leadership in any given area were able to interpret policies and modify directives so that they were in accord with the needs of the various microregions within their jurisdiction. Until the late 1970s, Taitou and its surrounding area was on the far southern periphery of its prefecture, far from the core city of Weifang. It was both difficult and impractical for Weifang factories to make contracts out in the southern boondocks when there were villages much closer to hand. Changwei prefecture apparently settled on a policy that defined the present Huangdao district as a basic grain area, while encouraging diversification nearer to home. It even went counter to Cultural Revolution policies in the range and extent of sideline enterprises it developed in the northern sector, but it did not attempt to spread these to the southern periphery.

The transfer of administrative control to Qingdao municipality did not make for any immediate change. The shift was necessitated by the location of the new oil port at Huangdao and the need to coordinate its administration with that of Qingdao's commercial and military harbors. The inclusion of Huangdao's surrounding rural area was incidental, linking it to a city that for almost three decades had met its needs for vegetable produce, meat and poultry, agricultural materials, and extra labor by turning to villages in its eastern suburbs or north up the railway line. In 1979, no one representing Huangdao district had been brought into the Municipal "Revolutionary Committee" (which was still running the city as late as the summer of 1980), and Qingdao's concern with its newly acquired territory seemed to be limited to the oil port and some fishing brigades.

Ferry services across the bay had still not been fully restored: there were several ferries whose use was limited to Huangdao oil-port personnel and their families. Anyone else had to rely on the one ferry a day which left in mid-morning from Huangdao and returned in the early afternoon. It was a passenger boat, not one on which peasants could carry goods and livestock to market. If one had complicated business to transact in town it would be necessary to stay overnight and catch the return boat the following day. In summer of 1980 I was told that a second run would be offered on the ferry service so that people could return in the late afternoon.

The general impression I received was that the city administration was not directly concerned with its new rural suburbs and was leaving the planning to the cadres at the district level, who at that time were appointees from the ranks of Cultural Revolution cadres. Like their Qingdao counterparts they were primarily concerned with the development of the oil port and the fisheries. The highest-ranking district cadre I interviewed assured me that Huangdao was carrying out party policy to the letter, explaining that in areas like this, which had a high population relative to land, it was important to "take grain as the key link," i.e., to concentrate on raising grain productivity and increasing delivery to the state. At the time, the press (and party policy) was recommending that surburban brigades with high population pay attention to vegetable and agricultural sideline production and develop industrial sidelines in conjunction with urban enterprises.

One expects that things have changed in Huangdao since then. By the spring of 1981, Shandong completed the process of disbanding the Revolutionary Committees of the Cultural Revolution, replacing them with more representative elected Peoples Congresses. The Cultural Revolution, which had taken a particularly leftist line in Shandong, was over in fact as well as in name. These changes in administrative leadership for Huangdao district, for Qingdao municipality, and for the province as a whole should bring economic policies more in line with those being formulated in Beijing. There is evidence that this process has begun: in 1980–81, several of Shandong's poorer counties and districts were allowed to shift some of their grainlands back into cotton and oil-bearing crops. More attention was being given to vegetable production, poultry and livestock, fruit orchards, and reforestation. Small collective or private-service industries were reemerging in the cities, easing the unemployment problem. Peasant efforts to develop new sidelines or participate in private sideline activities were being encouraged, or at least were not forbidden. New systems of work organization were introduced. By the end of 1981, virtually all rural units were participating in some form of the "responsibility systems" discussed in detail elsewhere in this volume. Taitou had indicated, in the summer of 1980, that it would institute a system of assigning output quotas for work groups within the teams and calculating renumeration according to output, a system which they felt would maintain and probably increase agricultural production while reducing work hours.

It is reasonable to assume that by now they have moved on to the household responsibility system, and reasonable to expect that some of Taitou's land will also be restored to commercial crops like cotton, soybeans, and peanuts—crops successfully grown in the past. And if the demand for contract workers lessens, it is possible the new district administration will recognize the need to encourage and assist the development of new sidelines and industries for the communes, brigades, and teams. The pace of development in Taitou, in Xingan Commune, and in the present-day Huangdao district seems to be related to the policies coming from higher up, that is, from the provincial and prefectural level, from the administrators of Jiao county (and later Jiaonan county), and most recently Qingdao municipality. Some of these had negative effects. There was overemphasis on grain production coupled with lack of attention to or discouragement of agricultural sidelines, handicrafts, and small industrial ventures. From 1966 to 1976 household sidelines met with criticism. Cropping patterns were changed to meet the call for grain deliveries to the state in disregard of soil and weather conditions, availability of irrigation, and people's dietary preferences and nutritional needs. These negative developments did not stem from the decisions of the teams and brigades or from the supposed ignorance or backwardness of the peasants. The decision was not theirs to make: they were not even consulted, although the slogans and speeches of the Cultural Revolution praised the peasantry and exhorted everyone to learn from them. When the policies met with only limited success, or failed, the blame was placed not on those who had conceived them but on persons at the grass-roots level. Team and brigade leaders or "bad elements" were convenient scapegoats.

The criticisms of this period that appeared in the Chinese press in 1980–81 are harsh to the point of overlooking the real accomplishments that also occurred. While Taitou's experiences are more representative of what was occurring province-wide and nationally, there were exceptions such as Anqu county which demonstrate that bureaucratic controls can be beneficial when administrators are imaginative and willing to take risks. The policies put forward after 1980 seem designed to counter the limitations of bureaucratic control by returning most of the planning and decision making to the households, teams, and brigades and by substituting elected officials for political appointees at the next higher levels of commune and district or county. While further information is needed about recent developments in the area, it seems reasonable to be hopeful that the negative aspects of past policies have been corrected, and that we will be seeing a more rapid and balanced development of rural villages like Taitou over the coming decade.

## Postscript

As this volume was going to press, China's State Council announced that fourteen coastal cities would open special export processing zones as sites for foreign enterprises and joint ventures—Qingdao is one of those fourteen cities. In addition, an industrial and research zone will be opened at Huangdao, coupled with a modern tourist center at Xuejidiao. It is expected that by 1990 the population of Huangdao will be double the earlier projections given in this chapter, and there will be over a hundred factories and related research and training centers.

It is as yet unclear whether regular employment in these new areas will be open to holders of rural household registrations. However, sandwiched between these new development areas, it appears that Taitou's villagers will continue to be employed at least as temporary laborers in the construction of housing, hotels, factories, and office buildings as well as in expanding port and railway facilities. Taitou's life, then, cannot but be radically transformed, even if its residents fail to get all the benefits of regular urban residence.

For villagers who remain at greater remove from these modern development projects, many of the issues raised in this chapter about the allocation of bureaucratic resources may still remain. The power of the bureaucracy has been greatly reduced, but with many supplies still coming from state-owned companies distributing their products through bureaucratic channels, the issue of potential administrative biases that help one region over another should not be forgotten.

## Notes

1. Martin Yang, *A Chinese Village, Taitou, Shantung Province* (New York: Columbia University Press, 1945).

2. Funding for this research came from the Social Science Research Council. The research was facilitated by Shandong University and the Foreign Affairs Office of Shandong province, as well as by the assistance and cooperation of the Qingdao Foreign Affairs Office, its Huangdao branch office (cadres), and responsible cadres of Xingan Commune and Taitou Brigade. Thanks are also due to the Weifang Foreign Affairs Office and Shijiazhuang Brigade, which I initially visited for ten days in 1977 and returned to three times for shorter visits in 1979–80.

3. Women rarely worked the first work period of the day because of the press of household chores, and they rarely worked late afternoon for the same reasons. As a result, few could earn more than seven of the ten daily workpoints, nor did they work for extra credits.

# Index

# About the Authors

MARC BLECHER is Associate Professor of Government and Chair of the Committee on Third World Studies, Oberlin College. He is the co-author of *Micropolitics in Contemporary China* and has written numerous articles on rural Chinese political economy.

STEVEN B. BUTLER is a Fellow of the Institute of Current World Affairs with a Ph.D. degree from Columbia University. His major research interest has been rural administration in China, and he has completed a book-length manuscript on the subject.

NORMA DIAMOND, Professor in the Department of Anthropology at the University of Michigan, is author of *K'un Shen: A Taiwan Village*. She has two works in progress, *Rural Economic Development in China* and *Changes in Taitou Village: A Study of a Shandong Village*.

NICHOLAS R. LARDY is Associate Professor in the School of International Studies at the University of Washington. He is the author of *Economic Growth and Distribution in China, Agriculture in China's Modern Economic Development*, and (with Kenneth Lieberthal) *Chen Yun's Strategy for China's Development: A Non-Maoist Alternative*.

VICTOR NEE is Associate Professor of Sociology at the University of California, Santa Barbara. He has written or edited *The Cultural Revolution at Peking University, Longtime Californ': A Study of an American Chinatown* (with Brett de Bary), *China's Uninterrupted Revolution* (with James Peck), and *State and Society in Contemporary China* (with David Mozingo).

WILLIAM L. PARISH is Professor of Sociology and Director of the Center for Far Eastern Studies at the University of Chicago. His works include *Village and Family in Contemporary China* and *Urban Life in Contemporary China*, both with Martin K. Whyte.

MARK SELDEN is Professor of Sociology and History and an associate of the Fernand Braudel Center, State University of New York at Binghamton. His work includes *The Yenan Way in Revolutionary China, The People's Republic of China*, and (with Victor Lippit) an edited collection, *The Transition to Socialism in China*. Following field work in 1978 and 1980, he is completing a volume with Edward Friedman, Kay Johnson, and Paul Pickowicz on *Wugong: A Chinese Village in a Socialist State.*

JONATHAN UNGER is Assistant Professor of East Asian Cultures at the University of Kansas. He is the author of *Education Under Mao: Class and Competition in Canton Schools, 1960–1980* and co-author of *Chen Village: the Recent History of a Peasant Community in Mao's China.*

THOMAS WIENS, with a Ph.D. degree in Economics from Harvard University, is an economic consultant on contemporary China. His works include *Microeconomics of Peasant Economy: China, 1920–1940* and "The Limits to Agricultural Intensification: The Suzhou Experience," in *China Under the Four Modernizations, Part I*, edited by the Joint Economic Committee, U.S. Congress.

DAVID ZWEIG is an Assistant Professor of Political Science at the University of Waterloo, Ontario, Canada, and a Research Associate of the Centre for Modern East Asia, York University-University of Toronto. He is currently a Postdoctoral Fellow at Harvard University's John King Fairbank Center for East Asian Studies. He is writing a book on the radical agrarian policies of 1968–1978 in China and the local response, having previously published in *The China Quarterly, Asian Survey,* and *World Politics.*